The Passive Eye

PHILOSOPHY   POLITICAL THEORY   AESTHETICS

Judith Butler and Frederick M. Dolan

EDITORS

The Passive Eye

GAZE AND SUBJECTIVITY IN
BERKELEY (VIA BECKETT)

BRANKA ARSIĆ

STANFORD UNIVERSITY PRESS
STANFORD, CALIFORNIA
2003

Stanford University Press
Stanford, California
www.sup.edu

© 2003 by the Board of Trustees of the Leland Stanford Junior University. All rights reserved.

Originally published as *Pogled i subjektivnost: Neki aspekti Barklijeve teorije vidjenja.* (Belgrade: Belgrade Circle Press, 2000).

"Film" is reproduced from Samuel Beckett, *Collected Shorter Plays* (New York: Grove Press, 1984), by permission of Grove/Atlantic, Inc.

Library of Congress Cataloging-in-Publication Data
Arsić, Branka.
   [Pogled i subjektivnost : neki aspekti Barklijeve teorije vidjenja. English]
   The passive eye : gaze and subjectivity in Berkeley (via Beckett) / Branka Arsić
   p. cm.—(Atopia)
   Includes bibliographical references and index.
   ISBN 0-8047-4642-7 (alk. paper)—ISBN 0-8047-4643-5 (pbk. : alk. paper)
   1. Gaze. 2. Subjectivity. 3. Berkeley, George, 1685–1753. 4. Beckett, Samuel, 1906–, Film. I. Title. II. Atopia (Stanford, Calif.)
B846.A77   2003
121'.35—dc21   2002154090

Original printing 2003

Last figure below indicates year of this printing:
12 11 10 09 08 07 06 05 04 03

Printed in the United States of America on acid-free, archival-quality paper.

Designed and typeset at Stanford University Press in 10.5/13 Garamond.

Contents

Acknowledgments — ix

Preface — xi

1. The Passive Synthesis of Contemplation; or, The Mirror in the Cabinet of Wonder — 1

2. The Active Synthesis of Reflection; or, The Eye at the Bottom of a Wine Vat — 18

3. The Passive Synthesis of Exhaustion; or, The Things in the Midst of the Eye

    GOD — 49

    PERSON — 96

    EYE — 139

Appendix: Samuel Beckett, "Film" — 177
Notes — 183
Works Cited — 201
Index — 209

Acknowledgments

This book is the result of, among other things, agreements and disagreements, the conversations and silences of the love and friendship of many people. I am grateful to my professors and colleagues Jovan Arandjelovic, Radomir Konstantinovic and Obrad Savic for their encouragement and comments.

I owe special thanks to Peggy Kamuf for her unfailing support; to Helen Tartar for the strong backing she has always given to the project; to Bud Bynack for scrupulous attention to the manuscript, and to Tony Hicks for being such a wonderful editor.

I am grateful to the Belgrade Circle NGO and to the State University of New York Individual Developments Awards Program whose help enabled me to complete this project successfully. Thanks also to the Grove Press for so generously giving me rights to reprint Samuel Beckett's "Film."

From the very first moment I started working on this project I enjoyed the help of Dusan Djordjevic Mileusnic. He assisted me in tracking down certain titles, patiently checked the footnotes, and was always ready to listen to my arguments about Berkeley's philosophy; for all that I am most grateful. But beyond that I am thankful for his friendship, for convincing me that such a thing as friendship was possible after all. I am particularly indebted to David Wills who carefully checked the English version of the book and who was kind enough to reveal to me secrets of the language. And my greatest debt is to Judith Butler for all her support, help, and generosity.

Preface

Argumentum: Exposition, account, summary, plot outline, invented narrative.

The argumentation I attempt to elaborate in what follows is close to the exposition of an invented narrative. That is to say that it is not a systematic scholarly account of Berkeley's theory of vision and still less of Berkeley's philosophy. Rather, I have used (and therefore misused) a set of claims and utterances "belonging" to Berkeley in order to develop a narration of what I call "the passive synthesis of exhaustion or iconographic subjectivity." Iconographic subjectivity could be conceived of as a possible way out of the concept of subjectivity developed by Descartes and its strategy of subjectivation rooted in the active labor of self-appropriation and self-distancing. It could be conceived of as a radical deappropriation, as the absence of any work, as an unheard-of passivity incapable of distance, incapable of any subjectivized point of view. That is why iconographic subjectivity has to be related to the problem of the eye and the gaze.

If it is possible for there to be a subject that does not know distance or any determined point of view, what then would its relation to the world be, what does it see when it sees the "world," and how is the difference between interiority and exteriority to be established at all? All of Berkeley's theses on distance, depth, and surface have helped me to elaborate the concept of the passive eye as the eye of an iconographic, exhausted subjectivity. And even though Berkeley himself never used the term "iconographic subjectivity" or "the passive synthesis of exhaustion," or, for that matter, many other concepts I have relied on in the interpretation of certain aspects of his theory of vision, concepts like simulacrum, copy, "yesbody," unconscious God, and so on, I can nevertheless imagine that if one fine day he were to walk into my room asking for an explanation of my interpretation of some of his theses, I would be able to say: "Look, George, here is the thing, this is why I did it. . . . " And I can see us spending the night discussing his philosophy and, come the morning, he would leave my room joyful and convinced that he had not talked about anything other than iconographic subjectivity.

I am trying to say that my argumentation here is faithful to Berkeley's philosophy insofar as to be faithful to the other means (among other things) to accept his/her words and to take them to their extreme, even to the extent of accepting also their possible radicality, or indeed their madness. That was my aim here: to try to take some of Berkeley's theses to their extreme. It was my aim for at least two reasons. First, because it is a "substantially" philosophical gesture to take things to their extreme. One could say that philosophy is precisely that thinking that takes the "thought" of common sense to its extreme. What, for example, is Descartes's "evil demon" if not the trivial thought of common sense that senses can sometimes deceive us, taken to its extreme? That is why what separates common sense and philosophy is fear: common sense is afraid to go as far as philosophy goes.

Second, it seems to me that some of Berkeley's theses, precisely insofar as they are taken to their extreme, could provide a significant voice to the contemporary debate over subjectivity and its subjectivation. My effort to draw from Berkeley's philosophy some conclusions and concepts that were not explicitly formulated in it was supported by the philosophy of Gilles Deleuze. It could be shown that Deleuze's philosophy is totally permeated by the experience of British empiricism—that is, by pluralism. My readings of Berkeley's philosophies were, therefore, guided by Deleuze's thesis from *Dialogues* that "the essential thing, from the point of view of empiricism, is the noun *multiplicity*, which designates a set of lines or dimensions which are irreducible to one another." In that sense, Deleuze's philosophy functions as the background to the whole argumentation I have tried to develop in this book.

Which explains the strategy I have used. Very often I quote Deleuze without warning the reader in advance that it is no longer Berkeley who is speaking, but Deleuze (or, more correctly, that it is now Deleuze who is speaking through the voice of Bishop Berkeley). I do that not only because of the "hidden" harmony that I detect between their philosophies, but also for two other reasons. First, because sometimes both philosophers use the same concepts: assemblage, collection, fire, becoming animal, and so on. Second, because by omitting to warn the reader that the proper name of a voice that speaks has changed, I was trying to develop *one* argument in the multiplicity of its voices. However, for those who think that the exact same sentence cannot be said through a seductive clamor (that is to say harmony) of different voices, that it cannot be at the same time said by the voices of different philosophers, there is always the possibility of recourse to my notes, where each voice is carefully and formally distinguished.

This is even more so because the voices of diverse philosophers (but of course, especially those of Berkeley and Deleuze) are connected in my analyses through the gaze of Samuel Beckett's camera. The whole screenplay of Beckett's "Film" is guided by Berkeley's thesis "*esse est percipii.*" One could argue that everything Beckett tried to achieve in his "Film" (everything he wanted to "gain" by what he called "the angle of immunity," for example) constitutes a remarkable interpretation of Berkeley's theory of vision. (Which points to the relevance of Berkeley's theory for film and especially for the structure of what Deleuze called "the affection-image.") That is why each chapter dedicated to Berkeley's theory of vision begins with reference to Beckett's "Film," and that is why references to Beckett's "Film" serve less as "examples" than as the most important interpretations of Berkeley's theory of vision. The whole text of the Beckett screenplay is provided as an appendix to the book.

My first chapter, "The Passive Synthesis of Contemplation," is an outline of the relations between gaze, eye, and "subjectivity" as they are elaborated in the philosophy of Giordano Bruno. My aim in this chapter is to give some of the main features of one form of passive subjectivity so that it can be compared to Berkeley's. I do not seek to develop or analyze Bruno's philosophy—that is why it is only an outline. The second chapter, "The Active Synthesis of Reflection," is dedicated to Descartes's "Optics," to his effort to subvert passive subjectivity and introduce active, reflexive, laborious subjectivity, and to his way of seeing as the means of appropriation of the visible. The third chapter, "The Passive Synthesis of Exhaustion" (divided into three subchapters), is an analysis of Berkeley's "resistance" to Descartes's optics, which resulted in a different possibility of subjectivation, one that may be determined as "the anti-Cartesian" revolution that I call "the passive synthesis of exhaustion."

Commenting on his book on Kafka, Deleuze once said that while writing it, he was thinking about what kind of interpretation would make Kafka happy. This was my aim in writing this book: I wanted to give joy to Bishop Berkeley.

ONE

The Passive Synthesis of Contemplation; or, The Mirror in the Cabinet of Wonder

> How do you expect me to trap anybody when I am trapped? So it is a movement from the active to a regression into passivity, into means you can afford.
>
> LOUISE BOURGEOIS

Hidden Connections and Secrets

The universe is not the world—this is one of the main theses of Giordano Bruno's mannerism. The universe is infinite, the world is finite. The universe is an open, infinite field that embraces innumerable worlds in such a way that none of them is a "center" of the universe or in such way that each of them is a "center." The universe, therefore, has innumerable centers. It is a decentered universe in which the difference between center and periphery does not exist insofar as Sun, fire, earth, water, or star is each a luminous body, a world or a center—"those magnificent stars and luminous bodies which are so many inhabited worlds, great creatures and superlative divinities: those which seem to be, and are, innumerable worlds not very unlike that in which we find ourselves."[1] However, each world, being a center, is decentered within itself. Each world is composed of water, earth, and air, whose different relations form its different configurations, thus making of it a world multiplied within itself. The center "belongs to difference." There is no such thing as a limb of the Earth.

This open infinity, established by Giordano Bruno during lectures in Wittenberg that the Danish prince Hamlet could have attended, can be apprehended only by the "universal intellect," which is nothing other than this universe insofar as it "fills everything, illuminates the universe and directs nature to produce her various species suitably."[2] The universal intellect is at the same time the eye of the world, "because it sees both the inside and outside of all natural things," and "the differentiator," "since it never tires of distinguishing the forms."[3] Being both an infinite eye and the differentiator that differentiates the differences, the universal intellect can immediately see the whole of infinity. On the other hand, finite beings can see the universe only through an optic and linguistic figure that mirrors infinity—mirror and metaphor, for the universe is the constant translation of one world into another, a disorder of different elements connected by connections that contradict one another—a pure chaos. "There is in your primordial nature a chaos of elements and numbers, yet not without order and series . . . there are, as you may see, certain distinct intervals."[4] The interval is a distance between elements inscribed in every connection of elements, a disconnection inscribed in every connection, the wild spot of order, a moment of hesitation of motion, the very truth of the chaos. Intervals are the potencies of all differences, the vertigo of the universe. But, on the other hand, this chaos has its order ("a chaos . . . yet not without order"). There is a hidden order that organizes chaos into a chaotic order that can be discerned only by a power capable of translating and connecting the differences by preserving their difference.

Metaphor is such a power. It can connect what is distant, and it can separate the inseparable. It is the "royal figure" because it can reveal the hidden spots of sense within the confusion of infinite chaos. "Metaphor equals God"—this slogan of the thought that emerges in the intersection of the sixteenth and seventeenth centuries means precisely that like God (who is "the first principle and first cause," who is everywhere and who can connect everything), metaphor can transport one extreme to the extreme of another extreme. Metaphor can "see" the *correspondentia recondita*, the hidden collusion of the things that are so distant that no eye and no ear can ever discover their connections. Needless to say, this divine metaphor is a figure that rhetoricians do not use, for they believe that metaphor should be "interested" only in relations of sense within a language, that it should be in solidarity only with resemblance: It should connect only things that are similar to one another, thus dismissing the dissimilar. But the metaphor that connects only similarities is nothing but a trope among tropes: it is not "divine." The "royal" metaphor, the one that escapes the knowledge of rhetoricians, is the metaphor that can connect dissimilarities and differences without reducing them, connect what is in opposition by preserving the opposition itself. That is why this metaphor is the oppositional metaphor.[5]

"Forest," for example, is at the same time the metaphor of the multiplicity of the natural things in a world (for there are as many natural things as there are trunks in a huge mountain forest) and of the oneness of the universe that in its open infinity embraces innumerable and different things. Again: "forest" is both the metaphor of darkness and error (for someone in the forest sees only shadows) and of truth and light, insofar as the path to truth leads through darkness and error. "Dog" is both the metaphor of deduction and syllogism (insofar as a dog, like reason, follows only one path when he is very well trained) and of excursion, of going astray, of leaving the path. That is to say that every oppositional metaphor is ambiguous, which is why rhetoricians do not like them: they cannot convince because they produce uncertainty. An oppositional metaphor is the intersection of (at least) two irreducibly different finitudes. It does not reconcile differences, for that would mean to negate, hide, or "betray" the truth of the universe, that banquet at which all kinds of dishes are served: "There are the salads and main dishes, fruits and common victuals, hors d'oeuvres and spices, warm and cold, raw and cooked, food of aquatic and terrestrial origin, cultivated and wild, ripe and green, food for the healthy and for the ill, dishes for gourmets and dishes for the hungry, ones that are light and substantial, bland and salted, tart and sweet, bitter and mild."[6] An oppositional metaphor is, therefore, divine because it is a "symbolic festivity" at which all kinds of ambiguities and oppositions are celebrated: "contradictions and differences have appeared, suitable to the various stomachs and tastes."[7] An oppositional metaphor does not interpret differences. It is their image.

As the true reflection of the truth (which is to say, as the reflection of ambiguities, contradictions, dissimilarities, and differences) the oppositional metaphor has to be itself ambiguous. Metaphor, the "key to the universe," emerges as a demonic key: one that can lock the door by unlocking it. Truthfully manifesting itself through metaphor, the truth remains hidden. And this hidden truth will be re-produced time and again, like an echo of the universe, for the metaphor of the forest will lead to the metaphor of the dog, and from one metaphor there will emerge another, thus forming an open, infinite set of metaphors. Metaphor becomes the "enchanted mirror of the world."

Contemplation of Shadows and Anamorphosis

However, in the world of sixteenth-century mannerism, the mirror still does not have the "dignity" of the mirror of subjectivity. Brunelleschi's mirror, which mirrors and thus constitutes the subject's gaze, will use for philosophical purposes only the philosophy of the seventeenth century, a philosophy already possessed by the idea of subjectivity. Until then, until the emergence of Cartesianism, until the emergence of the complex game between optics and the metaphysics of subjectivity, the mirror will be not the mirror of the subject, but the mirror of the world. The chaotic infinity of the universe connects itself with itself according to the logic of infinite mirroring: everything is a mirror reflection of something else.

The first "image," or the first reflection, is composed by God within a monochromatic desert of nothingness as the reflection of his gaze, of his image:

> it seems to me that those who will not understand or affirm that the world and its parts are animated detract from the divine goodness and from the excellence of this great living being and simulacrum of the first principle; as if God were jealous of his image, as if the architect failed to love his own work—he of whom Plato remarks that he appreciated his creation for its resemblance to himself, for the reflection of himself he sees in it.[8]

Since God's gaze is infinite, the first image itself has to be infinite. The first image, or the universe, is therefore an infinite image composed of innumerable images, and it emerges simultaneously with the gaze of God as the image of that gaze. The universe is the mirror of God, or, to put it differently, it is the gaze of God that falls through itself into its own infinite image. Every world and every being in the universe is only an aspect, a finite mirror reflection of the infinite gaze. A mirror mirrors the mirror, and the whole universe is made of reflections and their shadows and then of wheels of mirrors and the shadows of those wheels, so that infinity resembles an "ancient Egyptian looking object, evidently highly magical"[9] in which there are mirrors made of astral bodies, of stars and planets (which mirror the light of the first monad), mirrors of plants, animals, stones, as well as mirrors of the zodiac signs: the Sun is the mirror of the Moon. This means: the mirror reflection, like every reflection, resembles what it mirrors, however in such a way that the resemblance does not refer to sameness, but to difference. Every mirror image says that it is irreducibly different from what it resembles.

This universe, therefore, is similar to Leonardo's "cabinet of mirrors" and its optical labyrinth, in which each mirror deforms what it mirrors by reflecting it. Both to the "infinite sea" of metaphor and to the infinite image of the mirror will be assigned the same task: to connect (through reflection) differences and contradictions by preserving their difference and opposition. That is why the labyrinth of mirrors (optical as well as metaphorical) cannot promise the transparence of the world. If every thing in the universe mirrors another thing by multiplying the differences, then each reflection, each being in the universe, is only a cipher, a secret or invisible code of the "essence" of the visible, of the first monad or God, who remains invisible: "He [who sees] sees the Amphitrite, the source of all numbers, of all species, the monad, the true essence of the being of all things; and if he does not see it in its own essence and absolute light, he sees it in its germination which is similar to it and is its image."[10] The finite eye can see only the image of the image, which resembles light and therefore is different from it. That is why metaphor and the mirror have at the same time divine and demonic power: they reveal the truth by making it invisible.

If the whole universe is the first infinite image of God, of God's gaze, then every finite being, every finite thing, has to be a finite image. The difference between the thing and the image disappears. By creating the universe as its own image in which everything is reflected in everything, God's gaze contemplates itself, and so collapses into one with itself. That it contemplates itself means that God as the first active potency, as the potency of his own act and creation, creates passive potency, its own image "through which it can exist." Passive potency is twofold: "Without taking active potency into consideration for the moment, I say that potency, in its passive sense . . . may be considered either relatively or absolutely."[11] When it is considered absolutely, it is the infinite image of active potency, its substratum. It is the infinite passive mirror of God's gaze that gives him back his own image at the very moment in which this image is created, so that "this passive potency corresponds so perfectly to active potency that one cannot exist in any way without the other, so that, if the power to make, produce and create has always existed, so, likewise, has the power to be made, produced and created, for one potency implies the other."[12] Or, to put it differently, one potency is the other insofar as God is not separated from himself, insofar as he is what he looks at. Employing metaphors dear to Bruno, one could say: everything unfolds as if the active gaze composed the passive screen that, in its passivity, was waiting for the gaze of God to collapse into it, to fall into its own truth. Truth and truthful vision belong only to the first monad, which contemplates itself in its own passive image. Jealous of his own image, God keeps it for himself. Only God sees the universe, or, to put it differently, only God can see infinity. Only God can have the "complete view of the whole horizon." Only God has the absolute identity that is nothing other than the infinite sum of differences.

However, everything appears different from the "perspective" of finite beings. They are passive only in a relative sense, insofar as they cannot become one with active potency, but can only be created by it. As Schelling put it in his interpretation of Bruno, they have only "relative identity."[13] The universe is therefore made of innumerable passive screens that mirror or contemplate one another, but can never contemplate infinity. If the first monad repeats itself in the second, if the universe repeats itself in nature, and so on, then with each repetition, the light of the first monad weakens, so that finite beings can see only its deformed receptacle. That is why Bruno speaks of shadows. With each repetition, the quantity of "darkness" increases, and finite beings, those finite passive mirror surfaces, can reflect and see only deformed shadows, the deformed echo-images of the first light. This does not mean that God's power weakens or that there is something that escapes God's eye. It means only that finite beings, precisely because they are finite, can see infinity only from their own finite point of view. They can see only one aspect of the whole horizon: "To take an example (crude as it is), you might imagine a voice which is entire inside the whole room, and in every part of it." But even though it is entire inside the whole room, the finite ear can hear it only from its own position, and the way it hears it varies according to its position. Shadows, the deformations of images, therefore, are not the effect of God's lack of power, but of the very nature of finite beings. "The finite is related to substantial reality in such a way that only when it is multiplied by its square [or infinite potency] is it equal to substance."[14] Only if it is infinitely multiplied can the finite contemplate the whole horizon.

Since this multiplication is impossible, since the finite being will never become the infinite God, finite beings can see only images that are deformations of other images: seen from the position of the finite being, nature is a deformation of the first monad. Or, ideas, which are the forms of things, are deformed in the finite mind capable only of creating shadows of ideas. "Finite" ideas are shadows of forms, and therefore their deformation: "The forms of deformed animals are beautiful in heaven. Non-luminous metals shine in their planets. Neither man, nor animals, nor metals are here as they are there."[15] Clouds seen in the sky are thus mirrors that deform the forms of reflected animal bodies, making them even more beautiful. In other words, what the finite being can see is the anamorphic field of deformations. The anamorphic image deforms by connecting *prospettiva accelerata* and *prospettiva rallentata*, or by "dispersing" the forms from their depth onto their surface. (This is the source of the "secret" of anamorphosis that Athanasius Kircher called *magia anamorphotica*.)[16] The connecting of *prospettiva accelerata* and *prospettiva rallentata* refers to the procedure that "negates" the experience of the "central" or "direct" perspective in which the spectator was supposed to apprehend his own gaze by "apprehending" the vantage point while remaining withdrawn from the image, not being visible in it and not being visible to it. To put it differently, central perspective was based on the idea that the visible is the effect of the self-reflection of the spectator's gaze, and not the reflection of the visible field. The gaze that does not see itself reflects itself into the vantage point and then returns back to itself as its own self-representation. However, the connection of *prospettiva accelerata* and *prospettiva rallentata* announces that in anamorphosis, the forms themselves are "projected" outside themselves. They deform themselves, as it were, by going out of themselves.

And by entering the eye of the spectator. In that sense, every visible thing is an eye-image that looks at the spectator and enters his eye: "as a result, active and passive items of interest pass out from the eyes and enter into the eyes. As the adage says, 'I do not know whose eyes make lambs tender for me.'"[17] The spectator's eye is a receptive surface-image into which another image, another visual species, enters: the eye becomes what it sees, what it looks at. "There are also other types of feelings which come through the eyes and immediately affect the body for some reason: sad expressions in other people make us sad and compassionate and sorry for obvious reasons."[18] Without mediation, the eye *becomes* a beautiful sight or a sight of "disgrace." It is a succession of different images. That is precisely the meaning of the absence of mediation: the images that the eye sees are not representations, not the effect of reflection performed by the mind that excludes itself from the image in order to be able to see the image. Or, the images (everything that exists) are representations insofar as they are not the "first monad," the manifestation of the absolute truth (to be the absolute truth, images would have to be infinite), but they are presentations insofar as the finite eye does not distance itself from them, but contemplates them. Images are presentations of the visible that penetrate the passive surface of the eye. The eye is, therefore, a contemplative eye, an eye that does not distance itself from the visible, but is the outcome of the becoming one of the visible and itself. Since the eye is always within the image, insofar as the image has penetrated it, the spectator cannot withdraw himself and re-present the image to himself. The eye, and therefore the spectator, becomes the image, always a different one. In this logic of vision, vision "bonds the mind to itself," so that the mind also becomes what the eye sees: "the spirit is bounded through vision," says Bruno. The contemplative mind, therefore, is the contemplative eye.

But if the mind is contemplative, if it is the passive eye, how, by what force or power, is the synthesis of the visible and the eye then performed? Or, to put it differently, if everything that exists is a passive mirror, a finite image, the passive surface of the eye, if the universe is made of eyes, so that the visible, as Bruno puts it, "passes out from the eyes and enters into the eyes," then what force causes this passing, what power causes this entering of an eye into another eye? This power is the power of the "universal intellect," of the "eye of the universe," the power of absolute activity that induces nature (the passive substratum) to create "her various species suitably," the power that is both inside and outside of the created, the power that animates the created from "within," internally. What is only relatively passive (what is passive in such a way that its passivity cannot be identified with the absolute activity of God) is internally, immanently animated by the universal eye. This is to say: the universe is immanently alive, thanks to the universal intellect, but every finite eye, every finite thing, because it has only "relative identity," is at the same time passive—immanently alive beings are passive. As if activity were fatigue, as if beings that are alive were tired. When an image passes out of one eye into another, this "passing" is moved by the "first differentiator" that differentiates and animates the image and performs the synthesis of image and eye. The active synthesis of images exists only as an effect of the labor of the "eye of the universe," which at the same time produces images, differentiates them, and connects them. The eye is not what performs the synthesis, insofar as it is the passive screen that becomes what it sees and, at the same time, the eye is what performs the synthesis, insofar as the "universal intellect" is its internal principle. This activity of passive contemplation by means of which the eye becomes what it sees we will, therefore, call "the passive synthesis of contemplation." Two important consequences follow.

First: the image that emerges as the outcome of the synthesis is neither the image that entered the eye nor the eye itself. It is a new image insofar as it now also includes within itself the image of the eye. (The same goes for the synthesis of any two images: the overlapping of two images produces a new image.) This "new" image resembles the images of which it is composed, but this resemblance does not refer to the sameness of similar images: the resemblance is an irreducible difference. There are no two same images in the universe, which is another reason why images are presentations, and not representations. "The differentiator" differentiates images by means of their overlap, so that the universe truly becomes the proliferation of differences. In that sense, the eye that becomes what it sees always becomes "something else," something different from itself, as well as something different from the image-eye that entered it. Images emerge and vanish into new images, and everything is visible only for a moment, as it were, and only to a particular eye: no two eyes can see the same image. Only God's gaze can catch all the images at the same time, which is why only his gaze sees the truth that is the production of differences.

Second: if the eye is always a different image, and if vision is what bonds the mind to itself so that the contemplative eye is the contemplative mind, then, by the same token, the contemplative mind is always a different image, always something else: "So, indeed, there are many things which stealthily pass through the eyes and capture and continuously intrude upon the spirit up to the point of the *death of the soul*, even though they do not cause as much awareness as do less significant things."[19] Vision, the images (visible species) that intrude into the eye, also intrude into the soul, and to such a degree that they can "kill the soul," which is to say, transform it completely and make of it a different soul. The contemplative mind bonded to the eye thus "suffers" its own constant transformation, its becoming different images or different minds.

The contemplative mind, therefore, is far from what is said to be modern subjectivity, the labor of unification that mediates differences and subsumes them under its own identity, thus re-producing and reaffirming its distance from the visible. The contemplative mind is still involved in the visible: it is a node of passive, contemplative synthesis, a contraction of the transformation of one image into another. This mind could be called a larval self, a local, relative, provisional, and temporary "identity" that can easily be blown apart by the intrusion of other images. As what does not have the power of unification, this mind is also not one. What we call the "spirit," or sometimes the "soul" (for at least in *On Magic*, the meaning of the term "soul" shifts), is a provisional set of different souls and spirits. Different images simultaneously enter the eye, thus making of it a "set" of different images and thus making of the spirit a "set" of different spirits: "There are also worst impressions which enter the soul and the body. . . . Nevertheless, they act very powerfully through various things which are in us, that is, through a multitude of spirits and souls."[20] Every soul is made of souls, every eye is made of eyes, and the whole universe is the infinite differentiation of differences. Every finite being in the universe is already a world made of resemblances, differences, contradictions, and oppositions. It is formed as an order of disorder and as a disorder that cannot see itself, that escapes itself through the production of differences, or that escapes itself as the tired motion of differences. Finite beings are formed according to the logic of the oppositional metaphor. That is why the oppositional metaphor is the mirror of the world.

Frenzies of the Gaze and the Betrayal of Resemblance

The finite mind can try to "appropriate" the truth only through a strange, inhuman upheaval: only by giving up the multiplicities of differences, by withdrawing itself from the body, from the senses, by trying to distance itself from the universe in order to be able to see it in its universality, in order to be able to see what is behind the mirror of metaphor.

> The result is that the dogs, as thoughts bent upon divine things, devour this Actaeon and make him dead to the vulgar, to the multitude, free him from the

snares of the perturbing senses and the fleshly prison of matter so that he no longer sees his Diana as through a glass or a window, but having thrown down the earthly walls he sees a complete view of the whole horizon."[21]

By passing through a death that makes him dead to difference (to the sensible), this mind will be resurrected as a spectral "pure mind," or as the eye of pure intelligence that finally sees everything without frames (or windows), without intersections of horizons, without multiplication of horizons and views: he sees a complete view of the whole horizon. He becomes God, or he is driven by the desire to become God, to see the world through the eyes of God. He is driven by the desire to see the gaze of God. He is the symptom and therefore the announcement of the arrival of Cartesianism: "And now he sees everything as one, not any longer through distinctions and numbers, according to the diversity of the senses, or as varied fissures are seen and apprehended in confusion."[22]

He is, therefore, like a hunter who chases resemblances and differences in order to absorb them, as Bruno would put it, in order to mediate them into a self-same identity. And that is precisely the problem implied by his effort. The fact that he no longer see distinctions, diversities, and fissures is the effect of unification, and not differentiation. It is the effect of the reduction of similarities into sameness, and not of seeing the irreducible differences between similarities. It is the effect of the "procedure" opposite to the strategy of God's gaze: God sees innumerable differences at the same time, and he sees them as differences. He is the one who differentiates. God sees the whole horizon as the horizon of differences, whereas this finite mind sees the whole horizon as the horizon of sameness. That is why he will never become God: his effort will be diagnosed as a mad effort and his gaze as a frenzied gaze. The activity of unification and identification, the insistence on sameness, will be denounced as frenzy, as a heroic frenzy insofar as its "ideal" is God, but as frenzy nevertheless. In other words, the emergence of modern subjectivity, or at least of Cartesian subjectivity seen through Bruno's spectacles, looks like madness. And this madness will end badly. The one who chases differences will be chased. The hunter will be caught in the trap of distinctions and fissures: "but in that divine and universal chase he comes to apprehend that it is himself who necessarily remains captured, absorbed and united."[23]

His absorption within what he is trying to absorb is the outcome of the perversion of passive contemplation. The frenzied mind is the mind that rebels against passive contemplation, the mind that rebels against its own nature, and that tries to transform passive contemplation into active contemplation: active insofar as the object of vision will be chosen by the intention of the gaze, contemplation insofar as the one who looks at it will be absorbed by it. Instead of becoming something else over and over again, this mind distances itself and focuses itself upon one object of its intention or gaze. This object becomes the whole world and now absorbs the frenzied mind into itself, a strategy similar to the "strategy" of love. That is why, for Bruno, love will become the privileged example of heroic frenzy:

> Here *he regards one object* to which he is turned by his intention. A *single visage pleases* him and *absorbs his* mind. In a *single beauty* he is delighted and pleased, and is said *to remain fixed upon it,* because the work of the intelligence is not an operation of motion, but one of rest. And from that beauty only does he conceive the *dart* which kills him; that is, which summons him to the ultimate end of perfection. *He burns by one flame only,* that is, he is sweetly consumed by a single love.[24]

He is consumed by what he has chosen and vanishes into the one and the same. What he madly takes to be the whole horizon is nothing other than a "point" on the line of the infinite horizon. That is the ruse of the infinite proliferation of differences, the ruse of the infinite gaze of God: through its active contemplation, the frenzied mind will be absorbed into sameness, frozen in a moment of differentiation. Instead of apprehending everything as one and united, it will be reduced to sameness, while the differentiation of differences simply continues. The final "revenge" of differentiation: instead of seeing everything, the frenzied mind will end up in blindness, blind, precisely, to differentiation.

In the fourth dialogue of the second part of *Heroic Frenzies,* Bruno distinguishes nine variations on the blindness of the frenzied eye, or "nine reasons for the ineptitude, disproportion, and deficiency of the human sight and apprehensive potency toward things divine."[25] We are going to classify and name those different types of blindness, which are the outcome of the "betrayal" of resemblance conceived of as difference and not as sameness, in the following way.

ALLEGORICAL BLINDNESS

This blindness emerges as the effect of the betrayal of resemblance conceived of as *analogia.* Analogy is not only a resemblance of finite things, but the resemblance of all resemblances, the resemblance of relations. It has the power to connect all things by preserving their difference and, vice versa, it has the power to differentiate sameness: analogy establishes the connection between "to see" and "being seen." He who is blind to analogy is, therefore, "absolutely" blind. He is blind to everything. He is always already blind. That is why Bruno will say that this blind man is "blind from birth." He who aspires only toward the unity of divine things, thus "humiliating" the senses and differences, thus humiliating the eyes, will be "debased and humiliated by nature." He will be excluded from the world, sentenced to absolute darkness. He will be reduced to sameness to such a degree that he will not have the "divine image" even in his fantasy. He who wanted the disconnection among and negation of all differences will now suffer the absolute disconnection between "to see" and to "be seen": "to be seen and yet not to see the light, like a mole I came forth into the world to be a useless burden to the world."[26] His will for unification is now faced with inverted form, as if nature had said to him: you wanted to see sameness, now you will be sameness, you will see nothing, you will only be seen. He is, therefore, reduced to pure visibility. He becomes a thing, a visible object that does not see. He is simply exposed to the eyes of those who still see. This blindness, therefore, should be understood as a warning. It sends a message to all who see it: if you give up the diversity of the sensible, you will become a mole, a visible being that cannot see. In other words, this blindness is the figure that mirrors every blindness. That is why we call it "allegorical."

BLINDNESS OF JEALOUSY OR APPROPRIATION

This blindness emerges as the effect of the betrayal of resemblance conceived of as *consonantia*. Differences exist in a harmony established by "disinterestedness," and thanks to the labor of *consonantia*, even extreme differences "live" in peace with one another. *Consonantia* enables the "connection" between love and hate, day and night, water and earth. However, in order to be able to see those differences, the eye has to look at them with a paradoxical gaze (or paradoxical love): with a disinterested gaze (disinterested love), a contemplative gaze. The eye will see visible differences only if it manages to look at them without becoming interested in them. Or, the eye will see the visible only if it preserves the *consonantia* between vision and desire for the visible:

> Now, the word sight has two meanings. It can mean the visual potency, that is the power of seeing of the intellect or of the eye; or it can also mean the visual act, the application which the eye or the intellect makes upon the material or intellectual object. Thus when the thoughts are advised to curb the sight, it is not to be understood in the first way, but in the second because it is the visual potency become act which begets the *affection of the appetite*, whether sensitive or intellectual.[27]

Consonancy vanishes when the gaze becomes interested in the object of vision. The affection of the appetite focuses the gaze only upon the one thing that the eye desires, and that thing becomes the whole horizon. Or, as Bruno would put it, the eyes of the viewer become so "infected" by his desire for the appropriation of the visible object that it burns his sight by its light. He is then bound to the sameness of the desired object to such a degree that it "buries him . . . thus making him so hidden from himself as the light of his eyes is now hidden from him."[28]

BLINDNESS OF PERIPETY

This blindness emerges as the effect of the betrayal of resemblance conceived of as *copula*. *Copula* establishes relations between differences, but preserves their distance from one another. The innumerable finite things in the universe are connected by the labor of the *copula* insofar as its "logic" is not "either-or" but "and-and." One could even say that the work of the *copula* determines the way in which the infinite gaze connects itself with itself. The gaze that is not possessed by frenzy follows the logic of the *copula*. It is accustomed to the contemplation of ordinary things, and it "goes" from light to darkness and from darkness into the light again, from one image to another, from one mirror to another, even though those images could be in opposition or contradiction with one another. The light seen by the finite eye is therefore always a "shadowed" light, the darkened light of the lightened darkness. However, the frenzied eye is guided by a desire to see pure light and to see it immediately, through an absolute, intuitive gaze that would see infinity in a "flash," as it were, and not as an infinite succession of shadows. That is why it stops following the slow motion of the *copula*. And what is more, it arrests its own motion. It steps out of time and life. It deadens itself ("Why do I a dead man go wandering through the world?") in trying to focus itself and to catch the sight of pure light. But precisely thanks to this act, which will bring it to great "success"—the manifestation of pure light—the eye will be thrown back into the deepest possible crisis: the peripety emerges. All of a sudden there is a sudden twist. Through the elegance of a miraculous surprise that mannerism celebrates as a "shocking effect of the unexpected turn,"[29] the eye will be exposed to the divine light: "accustomed to contemplating ordinary beauties suddenly he was presented with one celestial beauty, a divine sun."[30] But, of course, the finite eye cannot endure the infinite intensity of the absolute light of the absolute. This light, therefore, will burn his eyes: "As a result his sight was destroyed and extinguished was the twofold light which illumines the prow of his soul (for the eyes are like to lighthouses guiding the ship)."[31] His ship is no longer guided. His deadened body floats adrift, instead of being launched into the motion of the *copula*: he is condemned to darkness. The light becomes the absolutely dark light—*lumen opacitum*.

BLINDNESS OF INSENSITIVITY

This blindness is the effect of the betrayal of resemblance conceived of as *conjunctio*. *Conjunctio* has a power similar to the power of *copula* insofar as it can connect different images of "small" intensity into another image, but in such a way that this "new" image still makes visible those "smaller" ones ("lesser splendors," says Bruno). Thanks to this similarity between *conjunctio* and *copula*, the blindness of insensitivity is similar to the blindness of peripety: "The fourth blind man exposes the reason for his blindness, a reason similar, though not identical with the preceding one."[32] Or, more precisely, the reason for this blindness is similar to the reasons of all types of blindness, for this blindness, too, is caused by the desire for the one and the same. It is the outcome of the "contemplation of the unity which removes him from the multitude," so that he, the frenzied one, "prefers that all objects be hidden from him for they could only annoy him by turning his view from that object alone which he desires to contemplate."[33] And, more specifically, the reason for this blindness is similar to that for the blindness of peripety insofar as it, too, is caused by the desire of the viewer to see the truth of the absolute light. However, the difference between them resides in their strategies. Whereas the "third blind man" wanted to see the absolute light immediately, the "fourth blind man" looks at the object that emits the light of smaller intensity, which enables him to look at it repeatedly, time and again: "This blind man did not suddenly find himself beneath the ray of light; it is for having gazed upon it too often or for having fixed his eyes upon it too much, that he has ceased to be aware of any other light."[34] He is patient, he does not want to be surprised, and he hopes that by looking at the same thing continuously he will finally see the truth of the light (the truth itself). However, by repeatedly looking at the same thing, he will become blind (insensitive) not only to all other visible things, but also to the motion of the "small differences" of which that very thing is composed: "So do I remain with spirit all intent upon the most living light which illumines the world and I am insensible toward all lesser splendors."[35] In other words, he will see only a rude "outline" of the thing he is looking at until he gets so accustomed to that "outline" that he loses sight of it, too: "And he says that the same thing happened to his sense of sight that happened to his sense of hearing; for they who have accustomed their ears to great uproars do not hear minor noises."[36] The blindness of insensitivity is, therefore, the blindness caused by habit, which is the force of the reduction of differences, the power of unification.

BLINDNESS OF WEEPING

This blindness is the effect of the betrayal of resemblance conceived of as *similitudo*. In every finite thing, in every ingredient of the finite thing, s*imilitudo* "discovers" the hidden traces of what is opposite to their nature and reveals the "secret" that any identity is possible only if it includes in itself what negates its nature. For example, the eye can see the light (everything visible has to be lighted) that is fire only if its retina is moist. Vision is the effect of the connection of water and fire. However, the eye that (like every eye) was once hurt by what it saw, but that (in contrast to all other nonfrenzied eyes) cannot overcome the pain and sadness caused by it, so that it constantly and excessively mourns and cries over what it has lost, is the eye that produces too much water, which then functions as an obstacle to the light (this eye overemphasizes one humor and thus excludes the other): "Because of the excessive weeping which has darkened his eyes he cannot extend their visual rays to the visible species and above all to that light again which in spite of himself and at the cost of his great pain, he once saw."[37] This frenzied blind man focuses and freezes his sight upon the image that has passed. He sees it again and again, and time and again he cries. His eyes, therefore, are no longer lighthouses (the intersection of fire and ocean), but founts or fountains (only the ocean): "My two eyes contain an ocean." Mourning, not joy, is the blind power of unification.

BLINDNESS OF SENTIMENTALISM

This blindness is the effect of the betrayal of resemblance conceived of as *aequalitas*. *Aequalitas* is the force that acts in every organism, and its function is to maintain the unity of its "diverse and contrary elements." It connects heart, kidneys, eyes, and soul within "one" organism so that, caught in this net of equilibrium, they all help and support one another, but nevertheless remain contrary to one another. This power also makes the different substances of which the eye is composed work together: the humid crystal of the eye has to be able to produce tears so that the fire of the visible species does not burn the retina. Non-excessive weeping is necessary for vision. However, when the eye finds that every image it sees is sad, that every image it sees hurts, that every image can cause weeping, and that the only pleasure it can take is pleasure in tears, it withdraws itself, unable to endure the differences, into the interiority of the viewer. It looks only at his internal cave, in which nothing from exteriority can reach him. Since he does not see anything else besides himself, nothing can hurt him anymore. In other words, he dries his eyes so that they are neither hot fire nor fountains, but dry sand: "Eyes not eyes; fountains no longer, you have poured out all the moisture which holds the body, the spirit and the soul together. And you, the crystal of the eye, which made so many external objects known to the soul, even you are consumed by my afflicted heart. Therefore, arid and blind I led my steps toward the dark infernal cavern."[38]

BLINDNESS OF SYMPATHY

This blindness is the effect of the betrayal of resemblances conceived of as *antipathia*. "The seventh reason allegorically contained in the complaint of the seventh blind man, derives from the fire of affection, from which some become impotent and incapable of apprehending the truth."[39] The universe is the simultaneous labor of sympathy and antipathy. Only thanks to antipathy is the universe saved from the excessive work of sympathy, which would assimilate all beings and crush the universe into eternal sameness: "This is why sympathy is compensated for by its twin, antipathy. Antipathy maintains the isolation of things and prevents their assimilation."[40] But when the fire of sympathy possesses the eye, then the eye assimilates all differences into one thing until its flame of sympathy burns it and transforms it into a dry dust.

BLINDNESS OF MASTERY

This blindness is the effect of the betrayal of resemblances conceived of as *amicitia*. All contradictory things in the universe are in a "friendly coalition" organized by God, who gives them different degrees of power. The contemplative eye is supposed to contemplate those different things without trying to "conquer" them. That is why the eye that, overwhelmed by passion, betrays *amicitia* and tries to "conquer" only one image will be "conquered" and "delivered" to blindness: "So does it occur that he who so gazes on high sometimes becomes overwhelmed by majesty."[41] The thing he loves will send the arrow of fire into his eyes, so that he will become the slave, instead of the master, the guided one, not the one who guides:

> Vile assault, cruel blow, unjust palm, acute point, devouring bait, strong sinew, bitter wound, pitiless ardor, harsh burden, arrow, fire and noose of the insolent god, who pierced my eyes, burned my heart, bound my soul and made me blind at one stroke, a lover and a slave, so that in my deep blindness every moment, everywhere and in every way I feel my wound, my fire and my noose.[42]

BLINDNESS OF FEAR

This blindness is the effect of the betrayal of resemblances conceived of as *concertus*. *Concertus* provides for the universal quarrel of all friendly things and beings, of one thing becoming the other thing through contemplation. But there is an eye that is afraid of entering the world. Such an eye closes itself before the images it longs for and sees only the sameness of its eyelids: "And therefore, he curbs his eyes from seeing what he most would desire and enjoy, as he holds his tongue from speaking to whom he most longs to speak, for fear that some defect of his glance or of his word might debase him, or in some way cause him disgrace."[43] Through its fear of unhappiness, this eye blinds itself, and thus happiness eludes it. By force of fear, this eye decides in favor of the mildness of sameness.

TWO

The Active Synthesis of Reflection; or, The Eye at the Bottom of a Wine Vat

> What I put on canvas might be described as representation, either representation or scheme.
>
> FRANK STELLA

Flame and Cat

Descartes waved goodbye to the Brunian universe. He announced, among other things, that clouds are not mirrors, but bodies made of "earthly" air through the evaporation of water from the earth. In addition to and "above" this cloudy atmosphere there is another element, the "first element," which in itself has nothing "elemental" (no elements of earth or of earthly air). This element moves with incredible speed, able in a moment to travel from the Sun down to us. The speed of its motion forms no shadow, and there is nothing dark in it, no spot. It is an eternal gleam, pure light or fire. Fire, which is another name for light, is the most refined element, one that moves in all directions, whose particles therefore tend to break the barrier formed by the clouds and to mix, while descending, with other elements, with earth, water, and crude air. This mingling composes different mixtures, called mixed bodies, made of different combinations of elements. However, light also fills the space between the bodies or within bodies, even "the gaps which are between the parts of the ordinary air we breathe,"[1] and in these pure pores of air creates bodies of an entirely different kind, absolutely "pure" luminous bodies, "so that these bodies, interlaced with one another, make up a mass which is so solid as any body can be,"[2] but which, nevertheless, differ from other bodies by not functioning as mixtures of different elements whose parts are opposed to one another in an eternal struggle of earth, air, and water. Luminous bodies exist in eternal harmony with themselves.

Luminous bodies also differ from mixed ones by virtue of their invisibility. The fissures and intervals between bodies are filled with pure bodies: with invisible, pure light. The reason for this invisibility is the nature of the motion of light and the magnitude of forms of the luminous bodies produced by this motion. Light moves at a speed faster than everything that is fast, and its parts (luminous bodies) are smaller than everything that is small, so that by its rapid motion they can go everywhere and enter everything. No body can act as an obstacle for them. Nothing can stop them. "For the form I have attributed to the first element consists in its parts moving so extremely rapidly and being so minute that there are no other bodies capable of stopping them."[3] The forms of pure bodies emerge through the rapid motion of light, through a motion so rapid that each form becomes nothing other than a rapid deformation. Form is therefore deformed in the very moment of its formation. As effects of this constant deformation, luminous bodies remain always without magnitude, shape, or position. They are pure flow, speed given over to itself. Formless, placeless, in all forms, in all places, everywhere and nowhere, such are these bodies of light. And for that reason they are invisible.

Such a body could be seen only by an eye that would itself be a luminous body, by an eye that could catch sight of the composition in the moment of its decomposing and that could follow the violent motion of pure light with no respect for organic contours, not stopping before them, but breaking them. In order for this rapid visibility to be seen, different eyes are needed, not human, but inhuman, divine, or animal eyes that instead of being turned toward slow, mixed bodies, are placed in a milieu that is fluid, rapid, vaporous, and formless. The pure visibility of the luminous body is promised to some other state of perception. "Earthly" and slow human eyes, which are themselves mixed bodies, cannot perceive these radiant points. They can see only what is formed and slow, only bodies composed of earth, air, and light. The visibility visible to human eyes, therefore, always requires elements of the earth. Slowness belongs to the earth. The motions of earthly particles are always slow, however fast they may be, because they remain heavy and closely tied. Slowness is what enables the subordination of intensity to extension and therefore the longer duration of forms and magnitudes, the positioning of forms. Other than pure light, other than the pure visibility that is invisible, all other bodies are visible insofar as they consist of a mixture of earth and light and insofar as the light in them is enveloped by slow elements that make it possible for the eye to see forms. Of course, these forms are in a process of constant deformation, but the speed of this process is not as great as that of the motion of the eye. The immobility of forms is nothing other than the slow motion of their deformation, which enables the eye to master forms or to see a form as one and the same. The form of luminous bodies is formed as an imperceptibly short interval between two motions, whereas in the case of mixed bodies, the motion itself is an imperceptibly short interval between two forms. There is a re-forming of forms and figures of mixed bodies, but the eye does not see them. For it, the figure remains immobile. That is why the figure or form is to be understood as an immobile sign of motion.

What is visible (mixed bodies) is the effect of a mixture of two invisibilities: of the invisibility of pure darkness (shaded, black earth, the absolute absence of light) and of the invisibility of pure visibility, of pure light. The visible is visible precisely because it is not pure visibility, pure light. Its visibility derives from the "covering" of the visibility of light with the nontransparent element of earth. And even though the heavy elements of earth cannot arrest the motion of light, they can slow down its speed and so maintain a form. The invisibility of light, therefore, is not excluded from the "space" of visible things. It resides within visible things themselves. Mixed bodies envelop luminous bodies, as if from within. They are, as it were, lanterns or chandeliers. In their interiority shines a sovereign white light, a pure, refined flame that makes them visible. The light that has "entered" nontransparent matter moves within it and becomes the force of its internal radiation. Visible and invisible come thus to be wrapped one within the other, maintaining themselves by their mutual "affirmation" and negation. In all visible bodies, there is the motion of light. All bodies are changeable, even though the speed of their change varies. The visible world is a glittering flow of forms, continuous change:

> I believe that countless different motions go on perpetually in the world. After noting the greatest of these (which bring about the days, months and years), I take note that terrestrial vapors constantly rise to the clouds and descend from them, that the air is forever agitated by the winds, that the sea is never at rest, that springs and rivers flow ceaselessly, that the strongest buildings eventually fall into decay, that plants and animals are always growing or decaying, in short, that there is nothing anywhere which is not changing.[4]

It is the difference in the speed and direction of the motion of particles of flame that causes the emergence of hot and cold, wet and dry, just as it brings about modification of the relief of mixed bodies, thus changing their appearance.

In a motion that spreads in all directions, light inclines toward what is not lightened and toward what is in darkness. Due to this inclination, it becomes clear that light is the condition of possibility not only for the visible, but also for vision. Bodies are seen not only because there is the motion of light in them, but also because this motion forces the eye to see them. The eye has to suffer the violence of this motion. "I would have you consider the light in bodies we call 'luminous' to be nothing other than a certain movement, or very rapid and lively action, which passes to our eyes through the mediation of the air and other transparent bodies."[5] Traveling through the air, the motion of light constitutes a straight line and comes to the eye along that line, thus composing in it an image of the visible. Light finds the eye through lively actions and then acts upon it, making of it the object of its rapid practice. The eye sees not because it looks, but because it is looked at. The object of vision has "found" the eye before the eye has "captured" it. The visual field does not sway because the eye is moving. On the contrary, the eye moves because it is forced into motion by the visible, which, before any gaze of the eye, looks at that eye and forces it to "establish" its gaze. Before it sees, the eye is seen by the gaze of things: it is the object of vision of its object of vision. Plants, stones, stars, objects, magnificent buildings and ruins, mountains and oceans, windowpanes and street lamps, all these allegedly blind traveling companions of human beings are in fact "spectators" who are already watching before any human eye in the world has opened to perceive that world. We are "beings who are looked at, in the spectacle of the world. . . . The spectacle of the world, in this sense, appears to us as all-seeing."[6]

In order to be able to see, the eye depends on being seen:

> so we must acknowledge that the objects of sight can be perceived not only by means of the action in them which is directed towards our eyes, but also by the action in our eyes which is directed towards them. Nevertheless, *because the latter action is nothing other than light*, we must note that it is found only in the eyes of those creatures which can see in the dark, such as cats, *whereas a man normally sees only through the action which comes from the objects.*[7]

Normally, man sees only by being seen. In order to become a body lit from the inside, or before it can become a visible body capable of seeing, the eye has to be the object of vision of some other gaze. Every gaze is conditioned by the gaze of the other and is produced by the gaze of the other. The gaze of every eye is at the "mercy" of the gaze of the other. The eye of a cat is the only exception to this rule. The cat occupies an ideal position, the point of intersection between the eye and the gaze: its eye is at the same time a luminous and a lighted object, light and darkness, simultaneously gaze and eye. The way a cat sees is based on a kind of a madness of temporality: it will be seen "before" it sees, and at the same time it will see the visible "before" the visible sees it. The cat's eye is, thus, a kind of scopic transfiguration, an absolute visual condensation: it is the spectacle of the world that is the eye, the intersection between the visible and the invisible, the ecstasy of the invisible into the visible and the visible into the invisible, neither visible nor invisible, but the archaic space of their common pulsation, the negated "bar" between light and shadow. The Renaissance image of the cat standing in darkness and holding the candle for Dante to shed light on his manuscript is replaced by a new, modern one: by the image of a cat whose eye is capable of seeing itself from inside, by the image of an eye capable of seeing its own gaze. The entire science of Descartes's optic will be moved by this fundamental desire—to appropriate for the human the power of the cat's eye, to enable the human eye to see its gaze. Modernity begins with a fantasy about *cat people*.

Map and Face

The world is all-seeing, and it *causes* our gaze. This production of the gaze occurs through a kind of "struggle" between mixed bodies. Bodies that enwrap the light of a stronger "inclination" have more power than other bodies to provoke the gaze or the eye, to make themselves be seen. "It is also good to note that when some object starts to act upon sensible organs stronger than the other and since they are not yet ready as they might be to receive their acting, the presence of this object becomes sufficient to make them, in that case, completely ready."[8] The action of an object upon the senses makes them sensible. Senses are indifferent to the sensible. The eye is indifferent to the visible, but the object of vision produces difference and induces in the eye the right "mood" for looking at it. Thanks to the "powerful" inclination of light, the object renders the eye capable of the gaze, as a result of which it can be seen by that gaze: "the action of that object will be able to render the eye momentarily capable to observe it."[9] The eye is forced to see. The eye has been obliged to see everything it has seen.

However, this provoking of the gaze no longer occurs, thanks to the contemplation of resemblance. Light is the motion that forcefully acts on the eye, but from the object to the eye, light transmits only light (luminous bodies), and not minute images of the object. Light can "paint" different images in the bottom of the eye by its action, but those images do not resemble the object. It is no longer a case of visible species traveling from the object to the eye, thus enabling the contemplative union of the visible with the eye. The image, conceived of as a resemblance, ceases to exist. Contemplation, as passive synthesis, as the falling of one image into another, is expelled beyond the limits of clear and distinct perception: "We must take care not to presume that in order to have sensory perceptions the soul must contemplate certain images transmitted by objects to the brain,"[10] not only because we could then no longer distinguish between eye and image, but above all because we would plunge into an endless proliferation of visual experiences that is both unknown to and invisible for itself. In order to see itself, contemplation would need "the third eye," which could observe the eye becoming what it looks at. But since there is no such eye, we have to avoid any proximity between the visible and the eye. We have to escape contemplation and take our distance from the visible. "Vision is not the metamorphosis of things themselves into the sight of them."[11] The eye is not going to see or be "populated" by ready-made images of objects. It is not going to be reached by anything from the object that resembles that object. Between the object that by its action upon the eye produces a sensation and the sensation itself there is now installed an irreducible difference: "the first point I want to draw your attention to is that there may be a difference between the sensation we have of light . . . and what it is in the objects that produces this sensation within us."[12] There is a difference between what we see and the visible. What we see does not resemble the object.

What exists between visual sensation and the visible is the relation between words and things: "Words, as you well know, bear no resemblance to the things they signify, and yet they make us think of those things, frequently even without our paying attention to the sound of the words or to their syllables."[13] Even though it is only a convention and not a resemblance that connects words and things, words nevertheless make things present to us by means of a conventionally established form of deciphering. The same goes for sensations. The sensation produced by the visible is conceived of as the word of the visible: "if words, which signify nothing except by human convention, suffice to make us think of things to which they bear no resemblance, then why could nature not also have established some sign which would make us have the sensation of light, even if the sign contained nothing in itself which is similar to this sensation?"[14] The visible does not copy itself upon the eye, it gives a sign of itself to the eye. This sign is different from the signs of "humane language" insofar as it is produced by nature, but it is similar to those signs insofar as its relation to the signified is arbitrary. By its origin, the visible sign of the visible is natural, but by its nature, it is "conventional." It is supposed to enable the reading of what has written it. In this sense, the visible becomes a network of signs inscribed in the eye: the visible becomes readable.

This visible language is now a veil that covers the visible itself. The visible remains underneath the signs through which it inscribes itself, concealed by a layer of arbitrary visibility—it falls into invisibility. Once again, the relation between the visible and the invisible is reversed. What was, as the essence of the visible, covered by visibility (pure light) now comes in turn to cover the visible by writing its text. The visible has given itself over to the signs that will represent it: instead of showing itself, it describes itself, narrates itself. On the surface of things there is a visible language functioning like a mask of things. "Little by little this unwanted and chatty visibility takes over the whole field of perception and opens it up for a language that then replaces it."[15] Visibility dresses itself in a costume, the costume of its language, which conceals it.

The visible world and the eye remain at a distance. The eye will be close only to the sign of the visible.

> As regards the position, i.e. the orientation of each part of an object to our body, we perceive it by means of our eyes exactly as we do by means of our hands. Our knowledge of it does not depend on any image, nor on any action coming from the object, but solely on the position of the tiny parts of the brain where the nerves originate. For this position changes ever so slightly each time there is a change in the position of the limbs in which the nerves are embedded. Thus it is ordained by nature to enable the soul not only to know the place occupied by each part of the body it animates relative to all the others, but also to shift attention from these places to any of those lying on the straight lines which we can imagine to be drawn from the extremity of each part and extended to infinity.[16]

The sign is inscribed upon the eye by the straight lines of light, which project one point of the object into one point in the eye and which connect those projected points by means of their motion. The object of vision is signified by a sign made of points connected by lines or rays of light. Light transmits the drawing of the object. By the mediation of light, the visible draws itself upon the eye. It draws its forms: the visible sign is a drawing of the *form* of the object.

However, even this projected form is different from the form of the visible thing. A last stronghold of resemblance is demolished. Projection distorts all forms to the point of becoming completely different from the form that drew them. The only visible given that is at the disposal of the eye is this natural artifact, this drawing that is supposed to represent the visible precisely through its difference from it. That is why the drawing is the sign: "Engravings . . . often represent circles by ovals better than by other circles, squares by rhombuses better than by other squares, and similarly for other shapes. Thus it often happens that in order to be more perfect as an image and to represent an object better, an engraving ought not to resemble it."[17] The entire force of this natural art is invested in a drawing of the shape that is not the reflection of the shape of what is drawn. The drawing, therefore, is not only a representative of the visible, but at the same time the only visible "thing" that can suggest the existence of an absolutely invisible visibility. The sign of the object becomes an object that "speaks" about a distant and invisible object. This is the entire logic of the engravings to which Descartes refers: "You can see this in the case of engravings: consisting simply of a little ink placed here and there on a piece of paper, they represent to us forests, towns, people, and even battles and storms."[18] Visible ink, through its absolute difference from the town, represents the town for the eye. From now on, the eye will see only an extreme impoverishment of the visible, only the *draft* of the visible. The visible has placed itself in the imaginary.

Thus the following paradox arises: the sign that is given in its symbolic function as signifier is to be found also in the imaginary as the representation or the image of the signified. The visual sign possesses the symbolic function of evoking the signified without any resemblance to it, yet at the same time it is the appearance of the signified in its own image, in its truth. That is why this mad imaginary-symbolic sign necessarily appears in the very figure in which it appears: "It is characteristic that the first example of a sign given by the *Logique de Port-Royal* is not the word, nor the cry, nor the symbol, but the spatial and graphic representation—the drawing as map or picture."[19] The map is not simply an image, but a descriptive drawing insofar as "description," for the geography of the seventeenth century, ceases to signify the verbal power of words, ceases to be a rhetorical means, and obtains instead the meaning of the *sense* of the image drawn through inscribing. The word appears as the figure of the image formed through measuring: either through geographical-mathematical projections of the totality or through horographic projections of the details of this totality, so that the projection of a detail is simultaneously a projection of its relations with the totality and the projection of totality is simultaneously a projection of relations between its fragments. Every projection projects relations. It is a description of connections, and therefore not only a description of the visible, but also of the situation of the visible. In short, the map can appear as the first example of the sign because it is the effect of the same procedure through which the visible inscribes its drawing upon the eye: the "points" of the visible object are projected onto the corresponding points of the eye, just as points of space, through mathematical projection, are transferred to and connected on the map. Every map emerges as an imitation of the actions of light. Cartography is thus rendered as the most "sublime" procedure through which the world addresses us.

Through its projections, this natural cartography also inscribes upon the eye the magnitude of the distance between two or more points, thus rendering the relations of convergence or divergence between the lines that connect those points. It is a map of the measure and order regulating the points of the visible. The map ascertains the projection of relations that can be measured. What was impossible throughout the entire history of optics now becomes possible—the reduction of the visible to measurable magnitude. Once *morphe* and *schema* are all that is left of the visible, once the entirety of the visible is reduced to form and figure as the outcome of the positions of its points, once every visible body is reduced to a geometrical body, then the algebraic treatment of this geometrical visible is made possible. Through a kind of natural analytic geometry, the visible reduces itself to a rational scheme, to measurable relations of magnitude. The self-mapping of the visible unfolds as a natural analytico-geometrical activity of nature that, unconscious of itself, performs the work of increasing and reducing the lines between visible points. Of the entire visible world there remain only magnitudes, mere amounts that can be calculated, enumerated, and estimated. This mathematization of the world is performed by the world itself. The world is always already mathematized, and like a strange visitor, it enters the dark little caves of the fountains built of sensible organs: "External objects, which by their mere presence stimulate its sense organs [the sense organs of the body-machine] and thereby cause them to move in many different ways depending on how the parts of its brain are disposed, are like visitors who enter the grottos of these fountains and unwittingly cause the movements which take place before their eyes."[20] The entire natural work of geometrization is performed unconsciously and stored in the grottos of the senses, where it determines the organs' means of "behaving." And since it is the effect of nature itself, since it is God's painting, this natural cartography becomes not only the representation of the visible, but its truth, the most perfect visibility. And by the same token, any other "artificial" map assumes the status of the perfect, ideal image, which bears in itself the idea of the perfect painting, for it is depicted in accordance with the principles of universal and natural analytic geometry, a geometry that releases the visible from the veil of diverse qualities and offers it to us in the transparency and glory of its truth, shows it to us, finally, for what it is—a pointlike object.

That is why the map offers "royal" advice to every painter: release the painting from any personal artistry. Let the world be projected into its own scheme. Let it draw its own chart: "Why, then, do we not methodically produce perfect images of the world, arriving at a universal art purged of personal art, just as the universal language would free us of all the confused relationships that lurk in existent languages?"[21] There is no reason not to, since universal and impersonal painting already exists, and the only thing needed is to try to mime it, as copper engraving does. There is therefore nothing unusual about the fact that when Descartes speaks of painting still not released from personal art, he refers to copper engraving, rather than to painting, for copper engraving is the "painting" that is closest to the nature of the map. Carved with a sharp object, it is formed point by point, thus following the logic of geometrical projection. Copper engraving is the disciplined record of points, a carving of shapes that maintains a resemblance to its object ("it is only in respect of shape that there is any real resemblance"), but even this resemblance is distorted, "even this resemblance is very imperfect, since engravings represent to us bodies of varying relief and depth on a surface which is entirely flat."[22] Instead of bodies and as bodies, we have only projections of bodies—nothing but maps.

The face of the world is a map, and the projection of this map is threefold. The map of the visible is inscribed through the pressure by which the motion of light rays projects points on the bottom surface of the eye so that to every point of the visible there corresponds a point drawn in the eye. The angles at which light falls on the bottom of the eye determine the magnitude of a hole made in the thin tubes that connect it with the internal surface of the brain. Once this map has been inscribed, the tubes will take over the role of light rays and, point by point, project the map from the eye onto the internal surface of the brain. There will therefore be displayed on this reflecting surface a map of a map. And the procedure will be repeated: dark corridors of tubes lead from the internal surface of the brain to the pineal gland, corridors that fill the role of light rays. The different apertures of the tubes determine the strength of "micro-motions" of animal spirits, which move along the routes of this internal pipeline and thus enable the projection of the map from the surface of the brain onto the reflecting surface of the pineal gland. It is this map alone that will become the object of sight. We always see the map of a map of a map. The immediate object of sight is therefore the effect of this threefold mediation.

The immediate, first language spoken by nature, therefore, no longer has the form of the trope or the metaphor (as in Renaissance mannerism), but of geometrical projection: it is a regime of signs whose signifieds speak of themselves in a mediate and roundabout way. The first determination of language appears in the figure of indirect speech. The eye, which saw only projections of the visible, and not the visible itself, will relate that "story" of the visible to the instance where the visible was never seen (the screen of the brain), and animal spirits will afterward transmit their version of the story to the pineal gland—it is always a matter of the story of the story, and not the story of the visible: "Language is not content to go from a first party to a second party, from one who has seen to one who has not, but necessarily goes from a second party to a third party, neither of whom has seen. It is in this sense that language is the transmission of the world as order-word, not the communication of a sign as information. Language is a map."[23] Language does not move from what was seen to what was told. On the contrary, it is a network of signifiers that signify signifiers.

In order for these networks of signifiers to gain the "support" of a ground upon which they will be inscribed, in order for the maps to have their paper, a face is required: "The face is what gives the signifier substance."[24] Every "ground" or surface is "facialized" insofar as it changes its expression as an effect of receiving and emitting signs. "Each time we discover these two poles in something—reflecting surface and intensive micro-movements—we can say that this thing has been natural as a face . . . it has been envisaged or rather 'facialized,' and in turn it stares at us, it looks at us . . . even if it does not resemble a face."[25] The quiet surface of the eye, the inside of the brain, the external edge of the pineal gland—these are all faces or reflecting surfaces upon which new maps constituting new facial expressions are projected by the motion of light or animal spirits. Even though the reflexive face is not the only possible outcome of facialization, in the Cartesian "structure of vision," facialization always has the reflexive face as its effect. This means: on the immobile surface, motions form a shape or a figure that emerges through the translation of the expression of qualities into magnitudes through a process of quantification of qualities, thereby establishing a signifying regime in which the sign refers only and always to another sign.

However, quantification and schematization do not affect only "external" visible bodies. Much worse, everything, including feelings and passions, will be subjected to the work of this cartographic art. Everything will be reduced to a map, and everything will be facialized.

> Thus, just as a figure corresponding to that of the object ABC is traced on the internal surface of the brain according to the different ways in which tubes 2,4,6 are opened, likewise that figure is traced on the surface of the gland, according to the ways in which the spirits leave from points a,b,c. And note that *by figures* I mean not only things which somehow represent the position of the

edges and surfaces of objects, *but also anything* which, as I said above, can give the soul occasion to perceive movement, size, distance, colors, sounds, smells, and other such qualities. *And I also include* anything that can make the soul feel pleasure, pain, hunger, thirst, joy, sadness and other such *passions*.[26]

Let it be clear: in the final analysis, everything becomes the object of vision. Sensations and passions become figures, produced always in the same way—through the projection of points. Sound and pleasure are also "pointlike" objects. There are maps of joy and maps of sadness, maps of pain and maps of thirst. Finally, every difference between qualities disappears. Qualities are differentiated only in terms of quantities. The difference between a red color and a book, between an animal and a city, becomes the difference in the number of projected points and in the form obtained through their connection. But all of these maps amount to nothing more than different pages of one and the same atlas.

From this point of view, it is understandable that only one passion can be fundamental, for all passions, including five of the "original" six, have themselves originated as the outcome of motions of animal spirits inscribed on the screen of the pineal gland. These faces, these geometrical portraits of bodies, feelings, and sensations, need to be deciphered. They have to be read. Immediacy of sensations is replaced by "immediacy" of understanding. A feeling can be the outcome only of reading and understanding: in order to feel love, it is necessary to decipher the map of love.

That is why every map provokes the passion of wonder—the passion of careful consideration. "Wonder is a sudden surprise of the soul which brings it to consider with attention the objects."[27] The effect of surprise refers to the nature of perception, to the eternal novelty of what is perceived. Perception is thus always the perception of the unknown: even when it perceives "the same," well-known thing, perception perceives it as different. Wonder is the only passion that can interpret perception because it is not the outcome of a motion. Nothing moves in the passion of wonder. No change occurs in the heart or blood, as is the case with other passions. That is why this passion is able to read projections. "The reason for this is that it has as its object not good or evil, but only knowledge of the thing that we wonder at. Hence it has no relation with the heart and blood, on which depends the whole well-being of our body."[28] Wonder is bloodless and bodiless passion. Wonder alone is not facialized. Wonder alone is something other than a map needing interpretation. One could say that wonder is the passion of all passions because it is the passion that escapes passion. Wonder is to be found in the place from which the visible can be read, in the place of understanding and consideration. Consideration of the wonderful, therefore, never deals with what is beautiful or ugly, with pleasure or horror, for such things are left to other passions. For its part, wonder considers only the essence of the wonderful produced by relations between points connected by straight lines.

The visible, therefore, is neither beautiful nor ugly, neither pleasant nor unpleasant. Mathematical projections, maps, and diagrams provoke only lifeless passion, a passion that does not affect the body, focused on consideration of the self-reduction of qualities of the visible to quantities, to nonqualitative magnitudes. From now on the visible has to be understood through precise consideration of proportions and relations and determined quietly and carefully, without recourse to bodily excitement.

But, if the visible has become the map drawn on the surface of a gland by force of some kind of internal geometrical secretion, if the visible has become a region of the cerebral cortex, this is because facialized projections have covered all ungeometricized, "wild," and "raw" visibility. This invisible, hidden visibility, irregular and disordered, falls into an anonymous existence. Such an "untamed" visibility cannot become a visible landscape. In the bestiality of its intense qualities, it is always and only a "scapeland," a visibility that escapes and runs, that cannot be inscribed in any chart, lying outside of all mapping, forever a terra incognita—a forest at the end of the world: "It used to be said that landscapes—*pagus*, these borderlands where matter offers itself up in a raw state before being tamed—were wild because they were in Northern Europe, always forests. *Foris*, outside."[29] The entirety of the visible covered by maps is left to its wild state, to Bacchanalias in the darkness of its forests. It is left to its formlessness, become wilderness unworthy of wonder. Yet we will never see these wild and unformed forests of visibility. To see means to focus on the waste regions of one's own cerebral cortex, to look only at one's own internal organs, to wonder at one's own pineal gland.

Blind Man and Wine Vat

We are always blind to the spectacles of the visible. We see only maps, but we see even them through a kind of a blindness of the eye. The visibility of maps is given through the "procedure" that constitutes the experience of the invisible. The experience of the visible is obtained by the gaze and its motion: the gaze that cuts the space of the visible and frames its fragments. However, in the Cartesian world, vision is formed by an opposite procedure. Here, the visible is offered through the absence of the motion of the gaze. It is given to an eye that is looked at, that does not look, to an eye without gaze, to a blind eye. The gaze will be "located" either in things that look without the eye or behind the eye. The eye itself will be forever freed from the gaze. It will become a kind of motionless excrescence "stitched" onto the face, a lifeless organ that blindly stares at the world. The eye no longer has the function of vision. On the contrary, the function of the eye is not to see, it is to be a still, glasslike surface on which reflection and refraction of light occurs. The eyes of a living man are, therefore, as dead as the eye of a dead ox or any other large dead animal. They act as windowpanes through which light is refracted in order to create an image map on the bottom of the eye. Thus, the function of the eyes is similar to the function of holes in a vintner's grape-crushing vat: "Consider a wine-vat . . . full to the brim with half-pressed grapes, in the bottom of which we have made one or two holes through which the unfermented wine can flow." The juice will flow in a straight line toward the open holes and into a barrel, touching the grapes and bypassing them, despite the obstacle they constitute, and "even though the bunches of grapes support each other." "In the same way, all the parts of the subtle matter in contact with the side of the sun facing us tend in a straight line towards our eyes at the very instant they are opened, without these parts impeding each other."[30]

The analogy is clear: the vat is the world, the grape juice is the light, the bunches of grapes are visible bodies, the holes are eyes, and behind the screen of the eyes will be positioned another gaze. Just as the pressure of the juice makes it fall in a straight line through the holes and into the barrel, so light penetrates the eye holes and falls onto the bottom of the eye, moving along paths made of light rays. Light rays are always and only lines: "You must think of the rays of light as nothing other than the lines along which this action tends."[31] Light rays are paths that transmit the action of the motion of the visible. But there are an infinite number of straight lines because every visible object is composed of an infinite number of points, and these cannot all be reflected at the bottom of the eye because the light is weakened by other motions that unfold along other lines with which it intersects. Only a limited number of the points through whose connection a projection of the shape of a body—a geometrical figure—can originate will therefore be inscribed upon the eye. The visual field thus becomes a measure of the relations and positions of these bodies—geometrical space.

Geometrical space, therefore, is not real space, but a projection of the real abstracted from real space. It is a self-projection of real space through which the visible represents itself to the eye by its own "natural geometry," and not the way in which the eye represents the visible to itself. That is why geometrical space, as it is reflected in the eye, appears as the imaginary of real space itself, just as the geometrical body belongs to the imaginary of the "real" body. And precisely this belonging to the imaginary will ensure that the geometrical projection or vision (which is not the image) performs the function of the image: "Vision is ordered according to a mode that may generally be called the function of images. This function is defined by a point-by-point correspondence of two unities in space. . . . That which is of the mode of the image in the field of vision is therefore reducible to the simple schema,"[32] which means that its function is to represent. Its function corresponds to that of Dürer's screen or Descartes's window placed in front of the eye of a dead ox: in both cases, it is a screen penetrated by straight lines, connecting points of the object with points on the canvas or the reflecting crystal of an eye that will emit the image to another screen.

The eye thus becomes a mere screen, a "cold medium" on which points projected by straight lines are inscribed whose intersections enable the motion of a body to be seen, its position to be determined. The eye itself does not see. It performs a function of a blind man's arm, and the light rays perform the function of the cane with which the blind man touches. But both cane and light ray function as a tensed thread that leads the blind man and saves him from going astray, an invisible straight line that enables him to reconstruct a visible space and to determine the positions and form of objects in it. If it is possible to explain vision by the example of a blind man, it is because, paradoxically, the light itself does not have to be visible in this geometrical space: "this thread has no need of light—all that is needed is a stretched thread. This is why the blind man would be able to follow all our demonstrations."[33]

The blind man sees by reconstructing geometrical space: touching in darkness, he moves from sphere to cube, from rhomboid to rectangular surface, seeing the entirety of the visible in its geometrical truth. This blind vision requires a rod that will transmit to the arm the pressure of the motion of the different objects it touches, enabling the distance of the object to be determined, as well as its form. It may seem to us that such vision is imprecise or uncertain, but that is only because we are not blind enough, because we do not see well enough: "But consider it in those born blind, who have made use of it [a stick] all their lives; with them, you will find, it is so perfect and so exact that one might almost say that they see with their hands, or that their stick is the organ of some sixth sense given to them in place of sight."[34]

Only the blind see perfectly. Perfect vision is blind because only the blind see forms and distances exclusively. Only they see nothing but perfect geometrical forms: they alone see the truth of the visible. From this it follows not only that the "Cartesian model of vision is modeled after the sense of touch,"[35] but that it is modeled after a "degenerate" touch, one that functions at a distance. The touch of the blind man in the Cartesian construction does not establish the immediacy of touch. Descartes's bearded blind man does not touch anything with his hand. He touches with a dead stick, with a wooden object, something that sticks into the object and transfers its motion to the hand. The blind man's touching is a distanced touching. It remains distanced from the touched, just as the eye remains distanced from the visible. Both touch and blindness are therefore subjected to blindness. Blindness is the operative notion of sight, as well as of touch. Or, they are both subjected to the blind seeing that unfolds through the mediation of a nonexistent sense, the sixth sense, an insensible and dead sense—the wooden stick, the straight line. Neither touching nor watching unfold through the mediation of living, sensible senses. In order to enter Descartes's perfect world of geometrical figures, blind men, the dead, oxen, and insensible senses are all necessary.

However, blindness can become the operative notion of sight only if the blind man touches with a twofold sixth sense, only if he uses two sticks. The blind can grasp the position of the object and its shape only by the intersection of two canes. One stick, one straight line, forever remains only a straight line. It cannot form a geometrical figure such as emerges only by the intersection of light rays. The same goes for the immobile screens that are the eyes. In each eye there is only one projection. Imaginary geometrical space is dissolved into two images that differ, thanks to the different positions of the screens of the eye and the different angles of the straight lines that fall on them, thereby constituting different figures. Exteriority is thus doubled in two different monocular projections. And only the animal spirits moving along the paths of tubes will project these two images into one on the internal surface of the brain. Everything still unfolds as if we were blind: "And as the blind man does not judge a body to be double although he touches it with his two hands, so too, when both our eyes are disposed in the manner required to direct our intention to one and the same place, they need only make us see a single object there, even though a picture of it is formed in each of our eyes."[36]

But the projection of two monocular images into one does not occur instantaneously. It requires an additional work of reflection and transformation. Connecting two monocular images into one does not involve simply adding them together. It involves subjecting them to a process of transfiguration, synthesizing them in a way that always loses something, so that the synthesized binocular image is something that was not inscribed in the initial monocular images. To make a binocular image means to make a *new* image. Images of quantity (monocular images) are thus exposed to a qualitative change that aims at establishing a new, binocular image, an image of the totality of the figure. The binocular image is the only one that should be grasped as the correct projection of the visible. It is quite different from the monocular images, which appear as apparitions, as phantoms of the visible, and which will disappear in the moment in which a binocular image is formed. Between binocular and monocular image there is a difference in order and in kind, an essential difference: "the binocular perception is not made up of two monocular perceptions surmounted: it is of another order."[37] It is not a phantom, it is not a "floating pre-thing," it is the real "*thing*."

At first, therefore, the visible does not appear in a precise projection, but as a phantom that will disappear in the moment of composition of the binocular, "true" image. This "true" image, however, is not constituted by the work of the eyes. The eyes remain passive, blind planes on which monocular engravings are inscribed. The binocular image will be constituted by the motion of animal spirits that will project both monocular images onto the screen of the brain, positioned obliquely in relation to those images. The constitution of the binocular image also comes about through the mediation of the function of the image, through the projection of points, but this time it is through a distorted projection that projects only points in which lines drawn from the points of two monocular projections intersect on a cerebral screen. The binocular image is therefore obtained by anamorphic perception of two monocular images. This perception occurs as the motion of animal spirits and as the tightening of the optic nerve, which enables the image to function, enabling the projection of the images into one geometrical point: "That which is of the mode of the image in the field of vision is therefore reducible to the simple schema that enables us to establish anamorphosis, that is to say, to the relation of an image, in so far as it is linked to a surface, with a certain point that we shall call the 'geometrical' point."[38] Of course, in Cartesian optics, this point will be mediated by another projection, and the anamorphic geometrical point will be subjected to a gaze governed by perspective. But in spite of the difference between two binocular images (between a binocular image that is the projection of two monocular images and a binocular image that is the projection of the binocular image), every binocular image is an outcome of the work of abstraction. Every binocular image is an abstract image. It is given as an abstraction of a "natural" abstraction insofar as every monocular image is already an abstraction from the sensible, its schematization. The binocular image, therefore, is a schematization of schemes, abstraction of abstractions. It is not a "live" picture of the motion of the visible. It is, rather, an abstracting of all these differences, a kind of frozen snapshot of the visible—a sign sublated to the essential feature of the signified. The binocular image is thus established as the image of the very essence of the visible. Abstraction, the "abstract painting," is the essence of the visible.

As a result of this schematization and unification of the visible, the gaze of the eye is rendered futile. In order for quantities and extensions to be "seen," it is not necessary to add the gaze to the eye. On the contrary, it is necessary to "free" the eye from the gaze. The eye has to be absolutely passive, dead and blind. The living eye that sees is always interested in something—the gaze always has its own logic of motion determined by its own desires. It sees differences, intensities of the visible. The dead eye, however, does not search for anything. It is a mere projective plane, the substance of the map, an extension within an extended world. The eye is placed among things and becomes the thing. "The eye's already there in things, it's part of the image, the image's visibility . . . the image itself is luminous or visible, and needs only a 'dark screen' to stop it tumbling around with other images, to stop its light diffusing, spreading in all directions, to reflect and refract the light. . . . The eye isn't the camera, it's the screen."[39] The eye does not move, does not see. It is a "dark screen," a thing. And precisely because it is without a gaze, because it has no interest in the visible, because it does not frame the visible in accordance with its own desire, because it does not act maliciously toward the visible by cutting it up according to its own will—the blind eye is the only good eye. The blind eye has no appetite, knows no greed, welcomes every projection equally, and has no preferences. That is why it is the only eye that can bless the visible. The blind eye will give its blessing to all projections, and projections will all continue to be projected to the point of the gaze in which the motion of projection will be arrested, to the point that will perform the function of an evil eye, to the point of perspective, to the geometrical point of seeing—to the *cogito*.

Monster and Mad Point

The blind man prehends with his sticks only what he can touch with them, only what is in front of him and what is within reach of his canes: a tiny field. He will "see" only one body. The entirety of visible being, for him, is reduced to the one form that his canes draw. There is no horizon, no visual environment. And precisely this reduction of the visible to one form is what makes the blind man completely blind—the blind man lacks a surplus: "an opening upon a system of beings, visible beings"[40] is not available to him. He does not see that besides what he sees, there is a visibility encompassing what he sees, a visibility that we see as what we do not see: "Around what I am looking at at a given moment is spread a horizon of things which are not seen, or which are even invisible,"[41] things that are felt or sensed as invisible, to which vision is subjected, and yet that through their invisibility structure the visible.

The blind man is not thrown into such a field. Around what he sees there is nothing invisible, and it is precisely this absence of the invisible from his visual field that makes him blind. The blind man is blind because he does not see the invisible, because he does not see the amorphous background of the visible. That is to say the blind does not see the blind spot, or else he is blind precisely to the function of the gaze—that he, like the Cartesian eye, escapes the function of the gaze. The function of the gaze is to "involve" the spectator in the visible, but to involve him in the figure of blankness: to make the spectator present in the image as a clearly depicted presence of an absence. The gaze itself is found in the place of the blind spot, visible as invisibility. This basic function of the gaze is what motivates the Cartesian blinding of the eye: the eye has to be blind in order to escape the blind spot, in order to escape the function of the gaze. The blind eye thus becomes the "edge" of the visible. The visible is in front of it, but the gaze is behind it, completely withdrawn from the image. The field of vision becomes a framed segment of the visible, without any surroundings: a *scenography*—a scene in perspective.

The binocular image produced by the anamorphic intersection of two monocular images will be projected from the brain to the surface of the pineal gland and "seen" by a perspective gaze. The point of reflection of the binocular image drawn on the brain is found right opposite its "central intersection," right opposite the point at which all the straight lines composing it intersect. In the image projected on the gland, the projection of "the central intersection" also constitutes the central, privileged point. Everything still unfolds according to the logic of spherical geometry: in both images, this point is privileged because it corresponds to the privileged status of the center of the circle that remains on the same plane with the circle as the image. However, this image projected on the surface of the gland is still a passive emitting surface, a screen with no gaze, but it is the screen that will finally be exposed to the gaze of a substance whose entire work is, after all, to think, to feel, to want, to affirm, and to perceive. It will be exposed to the gaze of the *res cogitans*: "It is the soul which sees, and not the eye; and it does not see directly, but only by means of the brain."[42]

The *res cogitans* sees only the screen image. It is thus the gaze that sees the image. Or, the *res cogitans* is the point of the gaze that is not in the eye, a point this time constituted through the logic of conical geometry. This means that the privileged point of the "circle" or of the screen image is "raised" in a straight vertical line to the top of a cone, from which, above the circle or the image, the totality of the image can be apprehended, together with its privileged point.

The *res cogitans* is the point from which the privileged point, the "central intersection," is seen. It is the point of view that views the privileged point. It is the perspective of the central intersection. In raising itself to a higher point, by distancing itself from itself, the privileged point, which is nothing other than the vantage point, becomes the point of view that views itself, a kind of self-reflection of the visible. As in the logic of painting's artificial perspective, where the point of the eye can be deduced from the vantage point, here, in this dioptrical construction, the *res cogitans* as point of view is also the outcome of the self-projection of the central intersection of the visible. The point of view—trivially understood as the point that corresponds to the eye of the spectator placed in front of the canvas—thus changes its meaning. It becomes the point of the subject. The point of the subject is a product of the self-projection of the visible, "deduced" from the visible. One could, therefore, say that the subject is an effect of perspective, in the same sense in which one also says that it is an effect of language.[43]

And yet, it is precisely this point of the subject that is the condition of possibility for perspective, and moreover the condition of possibility for the visible. The circular reasoning that obtains in the relation of the subject and God is thus repeated. Just as the existence of God was proved only if and when the *cogito* established itself without God's help, just as the *cogito* was the condition of possibility for its cause, so the visible is structured as the visible only by the gaze of the subject that is its outcome. The subject becomes "a painter" of what has painted it. The *res cogitans* becomes the condition of possibility for the visible. But this "painting" is framed—it is a field of vision cut out of the field of vision. Its lateral extension is limited by the width of the scene, which gives it, in turn, infinite vertical extension, infinite depth. The Cartesian visual field, because it is a system of projections, extends downward in such a way that the lines constituting its intersection, the lengths of its depth, are always subject to shortening, even when they are parallel. And thanks to this shortening of infinite depth, the infinity of depth becomes finite. The vantage point becomes exposed to the finite gaze.

By placing infinity within finitude, by negating infinity, the modern conception of the visual field is able to emerge. According to this conception, the absolute gaze, the gaze of God, is excluded from the image. If the vantage point can be seized from the perspective of what is finite, then it is precisely that finite perspective, the perspective of the spectator, that will become the absolute perspective, a sovereign perspective and the perspective of the sovereign. Yet the sovereign is also excluded from the image by the same procedure applied to the "shortening" of infinity. The same kind of shortening of the length of the depth will also occur "on this side" of infinity where the points of the visible intersect at a point "in front" of the gaze of the observer so that the visible cannot capture his gaze.

The negation of God's gaze will enable the observer, who is outside of the image, to become sole possessor of the image, sole owner of the world, the only sovereign. The subject exists as an outcome of this twofold shortening of infinite lines:

> The lines that run through the depth of the picture, on the other hand, are not complete; they all lack a segment of their trajectories. This gap is caused by the absence of the king—an absence that is an artifice on the part of the painter. But this artifice both conceals and indicates another vacancy which is, on the contrary, immediate: that of the painter and the spectator when they are looking at or composing the picture.[44]

The spectator withdraws into an invisible "spectatorium." The subject—and that is why he is the subject—is not the spectacle in the theater of the world. On the contrary, the world becomes a stage for the spectacles of his gaze. And even when the subject enters the stage of the world, he will remain hidden and invisible, one who sees without being seen: "Actors, taught not to let any embarrassment show on their faces, put on a mask. I will do the same. So far, I have been a spectator in this theatre which is the world, but I am now about to mount the stage, and I come forward masked. (*Larvatus pro deo*)."[45] The mask, also, has eyes, which here literally appear as holes, behind which there is a face with blind eyes, behind which there is, therefore, another mask that hides the spectator. Behind the mask is only another mask, only an empty screen of the face with its blind eyes. The one who comes forward masked is the one who puts a mask over the mask, thus trying to remain invisible, thus trying to escape the function of the gaze.

Escaping the function of the gaze through the blinding of the eye is a strategy to escape the blind spot, to establish an absolutely transparent visible. Of course: If the gaze functions as a blind spot, if it is the blind spot for itself, invisible to itself, then the constitution of the visible without the spot is supposed to accord the gaze its accessibility to itself, its becoming visible to itself. Escaping the function of the gaze thus emerges from a dream about the gaze that would not escape itself, that nothing of the visible would escape. The blind eye is an outcome of a dream about the absolute absence of blindness. And perspective, it seems, is the only means at the disposal of this dream, for "perspective provides a means of staging this capture (of the subject) and of playing it out in a reflective mode."[46]

Perspective is a staging of the capture of the gaze of the subject in the image, but in such a way that through the system of reflection of the vantage point onto the point of the subject, the subject can see that he is captured and so can see himself seeing, capture his gaze within his gaze. Through seizing itself as the privileged point, the gaze would capture itself, the gaze would look at itself and would escape its function of escaping, its function of being the blind spot for itself. But this absolute seeing without the blind spot is possible only under protection of a system of projections by means of which this "operation does not necessarily require that reference be made to any point exterior to the projective field that is the cone."[47] If projection of the intersection onto the viewpoint unfolds according to the logic of the projection of the center of the circle onto the top of the cone, this means that the split between subject and gaze can be negated only within the cone, within the "framed" visible, within a finite space, insofar as only in a finite space is it possible to fix a point as the point of the center of this finitude and therefore as a point that would reflect itself and look at itself. The entire logic of the "shortening of infinity," of placing infinity within finitude, was an attempt to fix the privileged point, which becomes privileged precisely through centering. But such a shortening is itself based on a blind spot:

> What is more we recognize that this world, that is, the whole universe of corporeal substance, has no limits to its extension.
> For no matter where we imagine the boundaries to be, there are always some indefinitely extended spaces beyond them, which we not only imagine but also perceive to be imaginable in a true fashion, that is, real. And it follows that these spaces contain corporeal substance which is indefinitely extended.[48]

Behind one finitude there are other finitudes. Every finitude is encompassed by an infinite multitude of other finitudes. Infinity appears as deprived of a center. And in infinity the vantage point becomes "infinite." There is no point that can provide for the constitution of the privileged point of the subject. In an open infinity, the point of the subject cannot exist—unless we accept the logic of the people of Lilliput: the universe can be infinite, but its center is, nevertheless, Mully Ully Gue, Emperor of Lilliput, "delight and terror of the universe . . . whose feet press down to the centre, and whose head strikes against the sun,"[49] which is an irrefutable truth proclaimed by Mully Ully and approved by the people of Lilliput. The Lilliputian truth about the center of infinity is, therefore, conventional. Mully Ully is an arbitrary center of infinity, an absolute subject established by convention.

This is precisely the procedure to which Cartesian optics has recourse in an attempt to save the point of the subject. Cartesian infinite space is obtained through analytic and "infinitist" geometry in which it is possible to turn toward the Lilliputian phantasm of arbitrary centering—analytical space is an infinite space with an arbitrary center established by the intersection of straight lines. However, this same procedure of analytic geometry applied to optics rests upon a paradox: the center of analytical space is determined arbitrarily, in contrast to the viewpoint in optical space, which is the outcome of the motion of the visible itself, its self-projections. The point of the viewpoint is thus the embodiment of an impossibility: it is an arbitrary-natural point. It is the arbitrariness of the natural and natural arbitrariness.

Establishing the viewpoint as an arbitrary-natural point, as a necessary convention, opens up the possibility of ascertaining the point of the subject, the exclusion of the subject from the image. But the blind spot of this procedure is that by producing the framed finite space in which the subject can see himself, perspective posits framed space as a perspective image within another space that the subject cannot see and in which he is caught. Escaping the function of the gaze, establishing the point of the subject through the projection of central perspective, is thus possible only through a blindness to the infinity of perspective that always already captures the subject: "Perspective posits a point 'encompassing' space within a space,"[50] a point that encompasses perspective or encompasses the point of the subject. This means that perspective acts as a kind of "*perspective meditation*" in which the subject encompasses itself and its exclusion from a perspective image, remaining blind to the logic of perspective meditation, blind to the fact that he is just a perspective image for another point of view. What escapes the subject is this invisible field of voyeuristic infinity that always already captures him.

But this blind spot is not just inscribed "behind the back" of the subject. It is inscribed in the subject itself, in its core. If the *res cogitans* is the gaze that is not projected in projections, if it is the objectifying gaze that cannot objectify itself, then it is the gaze that does not see itself seeing. By looking at the visible with no spot, by not inscribing itself as blind spot within the visible in order to see itself as blind spot, the *res cogitans* becomes something visible that does not see itself. It is blinded by itself. Clearly, this gaze of the *res cogitans* will become visible only through the production of another gaze that objectifies it. It is necessary, therefore, to go further or deeper into the interiority of this interiority to locate the thought of the gaze finally as the gaze of the gaze: "Descartes already sees that we always put a little man in man, that our objectifying view of our own body always obliges us to seek *still further inside* that *seeing man* we thought we had under our eyes."[51]

As a result of this strange self-reflexive ploy of the gaze, of this "metaphysical projection" projected in accordance with the logic of "perspective meditation," there emerges a point of view that views the *res cogitans*, the seeing of the gaze, a thought of vision. The geometrical point becomes a metaphysical point—the *cogito*. The *cogito* is a knot in which the subject of optics and the subject of meditations are tied together in such a way that they condition one another: the *cogito* is an effect of the self-reflection of the gaze, which itself is established in its visibility as an effect of the *cogito*. The *cogito* is thus supposed to become the gaze that sees itself. It finally comes close to being a cat's eye—it is structured as an eye that is the gaze, as the gaze that captures itself before it is captured.

The entire problem now is to avoid the logic of "perspective meditation." In order to escape the infinite production of the *cogito*, the production of the *cogito* of *cogito* and, at the same time, in order to establish a *cogito* that would not be blind to itself, the *cogito* has to become a split gaze that takes itself as its own "point of distance," objectifying itself by seeing itself in the very moment at which it distances itself from itself. The *cogito* is supposed to be established as the blind spot that sees through itself, as a paradoxically transparent blind spot. In other words it has to double itself.

The *cogito*, therefore, doubles itself into a self that is watched and a self that watches, a self that is passive and a self that is active, a self that is affected and a self that affects, to a You-self and to I-self. One *cogito* is always at least two *cogitos*. One point of view is always at least two points of view. It takes two for there to be a subject. According to "subject calculus," one does not exist. The subject is never "I." The subject is always "We." The *cogito* is always a couple. And this pair is a passionate pair, a "loving" *cogito* in which mediation appears in the figure of a domestic quarrel that produces difference. That is why the idea that the *cogito* could become the same to itself, that by escaping the function of the gaze it could become a bureaucratic self-identity, appears blind to the very nature of the *cogito*. The *cogito* always appears as difference and as the production of difference, as distance distant from itself, as distance that never diminishes. The *cogito* constitutively escapes itself. It can objectify itself only as an image in which it is no longer inscribed. The "now" of the gaze is necessarily inaccessible to it. The "now" of thinking is necessarily inaccessible to it, and no matter how much the thought-gaze is drawn into its interiority, from thought to thought, from homunculus to homunculus, there always appears, in this series of homunculi, in this proliferation of *cogitationes*, a *blinded* homunculus that does not see his own gaze, the gaze that is looking at him, standing behind his back.

The "madness" of the Cartesian attempt to introduce blindness in order to escape blindness, in order to produce an unblinded *cogito* that would see its gaze, whose gaze therefore would not be the blind spot, derives precisely from the fact that Descartes does not see that the blind spot is the condition of possibility for vision, the very core of subjectivation: blindness is structurally necessary. The "point of the subject" is nothing other than the blind spot. Descartes is blind to the insight that "what consciousness does not see it does not see for reasons of principle," that "what it does not see is what makes it see,"[52] that the *cogito* exists as *cogito* only insofar as it constantly turns around itself and in this turning misses itself, like the blind man who has not learned to use his canes. The Cartesian contrivance overlooks the fact that the arresting of the *cogito* in the endless motion of its failure to catch itself is actually the only way for the *cogito* to miss itself, insofar as its disposition is precisely its constant failure to see itself. Thus, the very nature of the *cogito* resists Descartes's attempt to invent a *cogito* that would not escape itself, whose gaze would become a captured transparency. The *cogito* appears as a force that insists on its own blindness. Instead of a *cogito* that sees through itself and sees itself, there appears a persistently blind *cogito* whose vision of itself is grounded in the constant escaping of the gaze or of thought, and whose nature, therefore, remains uncovered insofar as it is always encompassed by an invisible gaze. So the nature of the *cogito* appears to be that of a monster, and the monster appears to have the nature of the subject: "the monster is, constitutionally, he whose character cannot be revealed. He, like the modern subject in general, is located there where knowledge of him is omitted. His monstrosity is therefore structural, not accidental."[53] The *cogito* is located in the place of the gaze, and no matter how "deep" into interiority it withdraws, it will forever remain in the place of the gaze that does not see itself. The monstrosity of modern subjectivity resides precisely in its "structure," in the "fact" that it can be the subject only on condition that it is blind for itself, inaccessible to itself.

It was Pascal who first clearly realized that the *cogito* is a monster. It was Pascal who understood that the *cogito* is formed in the place of a gaze that does not see itself, that the subject is always blind to itself, and therefore that the point of the subject is necessarily a "mad point." The monstrousness of the subject's economy of subjectivation is precisely that the subject who sees through everything remains nontransparent to itself, that he who takes possession of the truth deceives himself about himself, is blinded: "His [man's] whole state depends on this imperceptible point [*point imperceptible*]," on the point of seeing at which he places himself, from which he sees everything, without seeing himself—his entire position, his "centering," is grounded on the impossibility of perceiving his own position:

You must not, then, reproach me for the want of reason in this doctrine, since I admit it to be without reason. But this foolishness is wiser than all the wisdom of men. . . . For without this what can we say that man is? His whole state depends on this imperceptible point. And how should it be perceived by his reason, since it is a thing against reason, and since reason, far from finding it out by her own ways, is averse to it when it is presented to her?[54]

The subject retreats from itself at the precise moment at which it is presented to itself because self-presenting is nothing other than self-doubling, departing from oneself in order to be positioned in front of oneself, against and opposite oneself. Doubling is the only possibility of the subject's self-representation, but it therefore implies the absolute impossibility of its self-representation. The subject never has a representation of itself, for the only representation it can have of itself is an image from which it is excluded, an image of someone else. The subject always considers he is someone else, like a beggar who thinks he is a king, like a madman who thinks his head is a pumpkin. His reason is thus placed in a mad position: the very nature of his reasonability escapes him, and the closer he gets to his truth, the more distant he is from it, according to the logic of the infinitely minute. The subject to whom the power of vision is given, to whom ownership of truth is assigned, who is posited in the place of the master, can see everything except himself. He is thus in the paradoxical position of a blind observer, of a "powerless master."

After all, the subject appears as a monster because he never becomes a powerful master, the transparent "blind spot." Everything always unfolds as in the experiment of the telescope and the boat: "if one looks through a glass directly at a boat disappearing into the distance, it is clear that the location of transparency [*le lieu du diaphane*] where one remarks the point toward which the ship moves will consistently rise, in continuous flux [*par un flux continuel*]." It is clear that the point of the subject, the point toward which the subject moves, will enlarge, becoming ever closer. The entire problem, however, is that the boat can move in this way endlessly, but that this point, even though it becomes larger and larger, "will never reach the one upon which falls the horizontal ray coming from the eye through the glass, but will continuously approach it without ever arriving at it, ceaselessly dividing the space remaining beneath this horizontal point without ever arriving there."[55] The vantage point will never coincide with the point of seeing. The subject will never coincide with itself. It will always remain a horizon for itself, approaching without ever reaching it. That is why the point of the subject can be denounced as a mad point—it is clear that the reasonable subject is constituted as a blind madman who is all the more blind insofar as he believes that he possesses the truth about himself, that he has become limpid reason, the pure "dentist's *cogito*": "The type of people that we shall define, using a conventional notation, as *dentists* are very confident about the order of the universe because they think that Mr. Descartes made manifest the laws and the procedures of limpid reason."[56] But their confidence is a grave error, for they do not realize that the dentist's *cogito* turns around itself in an attempt at self-synthesis, that it endlessly turns around its own blind gaze. The active *cogito* does not govern its gaze, but, quite the contrary, its gaze is its master. It captures it and makes it its slave: "Contrary to all appearances, and this is where the entire problem of the dialectic lies, it isn't, as Plato thinks, the master who rides the horse, that is, the slave, it's the other way round."[57] The blind gaze rides vision.

Good Painter and Evil *Cogito*

Of course, there is no escape from the visible, because it inscribes itself on the passive screen of the eye. The eye endures and suffers the visible. In order for the eye to cease being the object of vision of the visible, more is required than blindness, because the eye is always already blind. Conversely, for the visible to become blind, it is necessary that the visible itself disappear. Since this cannot be the case, since exteriority exists, it will continue to inscribe itself within the eye. It will leave its names and its figures there: "when an external sense organ is stimulated by an object, the *figure which it receives* is conveyed at one and the same moment to another part of the body."[58] Seal and wax. Thus the view of the *res cogitans* is made of figures, and even colors are reduced to figures. The visible is formed as a monochromatic paradise, mild and calm, deprived of irregularities. The entirety of the visible is petrified in the eternity of geometrical forms, in the eternity of names. The visible becomes a chimera, for the form is chimerical insofar as it results from the freezing of the motion of the visible, from an annihilation of the truth of the visible, which, in its invisibility, keeps moving and producing its truth, unstable and changeable: "What is real is the continual *change* of form: *form is only a snapshot view of a transition.*"[59]

What is real (transition, transience, deformation, the formlessness of the speed of a motion) is never to be seen. What is seen as real, this hardened, quiet form, is the abstraction of the real, the fiction of the real. Only one form that has originated through the condensation of the multitude of different, ungeometricized, irregular forms is seen. Only one "average" form of more forms is seen. It is always some kind of résumé of the visible that is seen, a mere forgery of the visible: "But, if several different figures are impressed in the one and the same place in the brain, with equal perfection, *which happens most often*, spirits shall receive something of the impression of every mentioned figure. . . . And precisely in this way chimeras and hypogryphs emerge in the imagination of men that dream awake,"[60] but also in men who do not dream awake, if between them a distinction can be made at all. All men dream awake, all men live in a dream.

It happens most often that like those who dream or who are insane, we see only chimeras, ghosts, and apparitions—motionless, frozen stills, a kind of average figure of all figures, a spectral essence of all chimerical forms. "When the successive images do not differ from each other too much, we consider them all as the waxing and waning of a single mean image. . . . And to this mean we really allude when we speak of the *essence* of a thing, or of the thing itself."[61] We allude to this photo-graphy, to this writing of imagery, to the figurative language through which the visible addresses us in indirect speech. Only now does it become clear: this language written on the "waxed" screen of the eye, because it inscribes only motionless, icy essences of a "flowing reality," can only be a language composed exclusively of nouns (photography is always a noun), for the noun announces the essence of the form, the essence of the visible.

The narration that, through the indirect appeal to the sensible, is inscribed upon the screen of the eye therefore has the structure of a film in which all the characters are points, straight lines, planes, triangles, squares, cubes, cones, or pyramids, all of them motionless snapshots following one another, and the muscles of the eye can "in a moment adjust the eye to all points of the object and thus make the soul see them clearly *one after the other*."[62] Motionless figures follow one after the other on the screen of the eye. One photograph replaces another. The *res cogitans* observes those immobile snapshots in their disconnected succession and then itself, through its own activity, connects them so as to make of their succession a motion of a totality—a motion picture. "The cinematographic character of our knowledge of things is due to the kaleidoscopic character of our adaptation to them."[63] We see totality where there is no totality. We read unconnected names as sentences and unconnected sentences as a single text. The *res cogitans* is at the same time like a director who creates an artificial world and the spectator who replaces "reality" with "the reality effect" of a movie, who considers a movie to be reality itself.

But it is not only the totality of a succession of snapshots that is the artificial product of the "artistic" work of the *res cogitans*. Every single shot is also an effect of it. In a certain sense, nature does not know much about "excellent painting." (For understandable reasons, the example of painting was closer to Descartes than was film.) Nature creates drawings through a natural geometry, through a natural reason that does not know of itself and that creates depictions that suffer various imperfections. Within every image that nature has made, different defects can be determined, as if some insufficiently wise painter had created them. That is why the *res cogitans*, wisest among the wise ones, which sees everything except itself, has to be dissatisfied by these images. It sees the imperfection of every natural image and takes it upon itself to perform the enormous, benevolent work of improving them, of retouching the photographs, of correcting and polishing the truth.

> But the intellect is like an excellent painter who is called upon to put the finishing touches to a bad picture sketched out by a young apprentice. It would be futile for him to employ the rules of his art in correcting the picture little by little, a bit here and a bit there, and in adding with his own hand all that is lacking in it, if, despite his best efforts, he could never remove every major fault, *since the drawing was badly sketched from the beginning*, the figures badly placed, and the proportions badly observed.[64]

Nature, no doubt, has made an effort: it has "rationalized" itself through a natural geometry and offered to reason a figural writing that was supposed to inscribe on the eye the essence of the visible by means of geometrical forms. And yet, this entire work was full of defects, the projections were quite badly conceived, figures in them were badly placed—nature has thus offered an untrue image of itself.

This paranoid perception of nature and of the senses as bad, unreasonable painters who have played the part of an evil demon in order to deceive reason belongs to the *res cogitans*. And that is who will correct their paintings. But in order to do that, it has to know that it is deceived. It has to know how the painting should be repaired so that it can become the image of truth. In order to retouch the painting of the visible in accordance with the invisible truth of the visible, the *res cogitans* has, therefore, to know this truth always already. Its original situation is, therefore: I look for the truth because I've already found it. The *res cogitans* possesses the truth of the visible before the visible appears in its visibility. The *res cogitans* is thus finally manifested as an evil eye, as the function of the evil eye. With its appetite for the truth, it will completely transform the visible and, what is more, condemn that visible to absolute catastrophe.

Why should the *res cogitans* be satisfied with mere retouching and correcting of bad paintings? Why should it be satisfied with adding a line here and a line there on the drawing that was badly sketched from the beginning? What is no good should be "wiped out with a sponge," says Eudoxus. It is best to abandon all depictions of nature and to "start all over again." It is best, therefore, to close the eyes, to remove the screens, and begin the creation of images of a new world drawn by wise reason, by its all-seeing blinded gaze. This is going to be a perfect world, finally, with no badly conceived images, in which every proportion will be assessed and measured correctly, a world of absolutely perfect figures, a world that cannot exist anywhere but in the eyes of the *res cogitans*. But even though it exists only "for" the *res cogitans*, this world is no less real. On the contrary, it is more real than anything real, because it is the only true world, one in which nothing is either dry or wet, neither light nor heavy, neither cold nor hot, in which everything sensible becomes insensible and supersensible—the world of perfect geometrical painting. So the *res cogitans* begins to create a perfect world that only it sees, established according to its own principles for the production of truth, a world in which the truth is nothing more than what was already posed as truth by the *res cogitans* itself. The *res cogitans* is thus overwhelmed by the madness of omnipotence, by the deception that the world it dreams awake is absolute truth, that it is the best of all possible worlds. It lives in a baroque scenography, in a baroque play that takes dream for the truth, that holds that the truth in its clearness can be prehended only in a dream insofar as it is only in dream that one sees clearly and distinctly: "Yes, and this is just the thing I saw before, as clearly and distinctly as I see it now, and it was all a dream."[65]

This insight of Segismundo, namely, that he only dreamed what he clearly saw, that he sees clearly when he dreams, and that he dreams even when he is awake, has its "mirror reflection" in Cartesian physiology. In a dream, fibers that connect the nerves and transmit the influences of external objects to the senses are loosened. The activity of exteriority is suspended, no longer able to pass to the brain. But, in turn, the imagination is awakened in a dream, Very different images invade the brain, images that memory has deformed, that imagination has imagined, and that compose the most amazing sights. Like any other image that comes from exteriority, dream images provoke strong motions of animal spirits on the screen of the brain, even stronger than that created by any map from the projections of the visible, "so that images that originated in a dream can be more distinct and more lively than those formed in a state of being awaken."[66] This is to say that they can also be more true, since clarity and distinctness are the criteria of truth. The true is what is prehended clearly and distinctly. The true is the dream, a dream world: "If it happens that the effect of some object that influences the senses manages to pass to the brain during the dream, it will not cause there the same representation it would create in the awakened state, but a different one, more noticeable and more sharp."[67] If it happens that during sleep exteriority breaks into the dream, its image is going to be corrected there. It will become a different image, more sharp and noticeable: the true image. Dreaming is, therefore, an ideal situation for "sharp perception," for the seeing through of the truth that is—the dream. Illusion, rejected in its untruth, is rendered as the force that overwhelms reason, as the very nature of reason. Reason, which was supposed to be free of fantasy and imagination, finally ends up as the pure labor of fantasy, as a facility for the production of illusions. The scenography produced in dreams, this illusive, dreamed scenography, is the only true world. Life becomes a dream.

THREE

The Passive Synthesis of Exhaustion; or, The Things in the Midst of the Eye

GOD

Wall

"Dead straight. No sidestreets or intersections." No *"sidestreets"* because every *"street"* is *"central,"* equally visible. No *"street"* is the privileged spectacle for the gaze that sees them all from all angles. The infinity of all possible spectacles is *"within"* an infinite eye supposed to be the eye of God. The gaze of this infinite eye is fast and dynamic. It does not know of an immobile point of view. It is a gaze for which each object of vision is a mobile point of view, a gaze that moves simultaneously in all directions, thus seeing everything at the same time. Nothing escapes this divine gaze. Every object of vision is immediately and completely visible. Every object of vision is exposed. This is, in short, the "structure" of Berkeley's "divine optics": the world is an infinite, flat surface on which everything becomes absolutely visible. Invisibility disappears.

Samuel Beckett visualized this surface as a wall in his screenplay "Film." In "Film," the object of vision, the character signified as O, slides along the flat surface of the world, along the flat surface of a wall, by covering himself, by protecting himself from divine perception, the gaze of the character signified as E.

However, "Until end of film O is perceived by E from behind and at an angle not exceeding 45°. Convention: O enters percipi *= experiences anguish of perceivedness, only when this angle is exceeded. E is therefore at pains, throughout pursuit, to keep within this 'angle of immunity.'"*

Beckett showed the world that Berkeley tried to prove. He showed that this surface is always within the mobile field of percipi *of God's eye. Because E sees the surface of the world at an angle of less than 45 degrees, E is never in front of the surface. The surface is therefore never a "deep" surface, but a flat surface that annihilates the very possibility of the angle of immunity. By insisting that O is perceived by E at an angle "not exceeding 45°" Beckett referred to the situation of the object of vision in Berkeley's world: if every object of vision is absolutely transparent, if, what is more,* esse est percipi, *the being of every object is to be perceived, then the situation of beings in Berkeley's world is intrinsically unbearable: "O, entering perceivedness, reacts (after just sufficient onward movement for his gait to be established) by halting and cringing aside towards wall." All our thoughts, gestures, wishes, desires, and secrets are always already exposed and visible. There is no joy of invisibility or secrecy.*

That is why Beckett tried to "construct" an object of vision that would do anything to escape this unbearable exposure: "O . . . hastening blindly along sidewalk, hugging the wall on his left, in opposite direction to all the others. Long dark overcoat . . . with collar up, hat pulled down over eyes, briefcase in left hand, right hand shielding exposed side of face." O wants the happiness of imperceptibility. He wants his being to escape the being of the Other. He wants to be left alone, if only for a moment. Of course, nothing like that happens. "E's searching eye, turning left from street to sidewalk, picks him up." O's being will be maintained time and again as a picture in the visual field of God.

Picture and Image

The Berkeleian person emerges in the moment of awakening of the Cartesian subject. What the Cartesian subject never saw—dark forests, the trembling of irregular forms, the mixture of horizons—composes the visual field of the Berkeleian person. Instead of the Cartesian visual field, which was a "glass garden" of universal laws, the Berkeleian person is located in a minute order that is itself situated in a jungle of visual lawlessness because the visual is here not geometrical. There is no such thing as *natural* geometry. The visible world never offers itself to the gaze as a world nicely formed into geometrical bodies by the system of projections. Maps and diagrams are the effect of the mathematical imaginary and are fantasies we have no reason to follow, for we do not see them.

We never see what geometricians see, and "everyone is himself the best judge of what he perceives, and what not."[1]

Just as we cannot say that we perceive geometrical projections because we do not perceive them, we cannot say that we perceive with the *help* of geometrical projections, since we do not have the experience of that help. And thinking that we have the experience of what we had never experienced means thinking like those who claim that they see geometrical bodies, even though they do not see them at all. Who ever saw a distant object as a point of intersection of two straight lines? No one:

> those lines and angles, by means whereof some men pretend to explain the perception of distance, are themselves not at all perceived. . . . I appeal to anyone's experience whether upon sight of an object he computes its distance by the bigness of the angle made by the meeting of two optic axes? Or whether he ever thinks of the greater or lesser divergency of the rays, which arrive from any point to his pupil?[2]

To claim, therefore, that the visual field is made of geometrical intersections is to claim that we see what is in itself invisible. Such claims cannot account for the nature of perception. They cannot offer an explanation of how the visual field is articulated. That is why the idea of geometrical vision ends up eliminating the proper objects of sight.

To assume, as Descartes did, that straight light rays engrave geometrical projections on the retina is to assume that the objects of sight are tangible inscriptions—that is to say, the proper objects of touch. Those projections draw maps that "are so far from being the proper objects of sight that they are not at all perceived thereby, being by nature altogether of the tangible kind, and apprehended by the imagination alone."[3] According to Berkeley's interpretation of this geometrical thesis, the tangible rays that touch the tangible eye, thus forming only something that is itself tangible for the encounter of two proper objects of touch, cannot produce the proper object of sight (the finger touches only something that is hard or soft, but it can never touch yellow or red.) That is why the understanding will have to imagine those tangible objects as visible objects, thus making of itself a kind of internal theater in which the transformation of the tangible into the visible is performed. This imaginative reason fantasizes that the maps it sees are the proper objects of sight, that they are pictures, thus remaining blind to the insight that they are images, not pictures. In accordance with Berkeley's terminology, which follows the trajectory of a distinction between proper objects of touch and proper objects of sight, geometrical projections can never be pictures, but always and only are images: "The pictures, so called, being formed by the radius pencils, after their above mentioned crossing and refraction, are not so truly pictures as images, or figures, or projections, *tangible figures projected by tangible rays on a tangible retina.*"[4] Geometrical projections, therefore, are not an unstable bundle of colors such as would constitute pictures, the proper objects of sight. Projections or images are, on the contrary, the invisible, tangible objects formed as the effect of the intersection of light rays on the retina, as took place with Descartes's blind man and the intersection of his canes. In the Cartesian world, therefore, the visible remained invisible.[5]

Berkeley claims that *resemblance* is the vital motive of Cartesian optics and that it brings about the elimination of the visible and a scandalous celebration of the power of the blind eye: "these tangible images on the retina have some resemblance unto tangible objects from which the rays go forth."[6] However, what is at stake here is not the resemblance between the projection and the projected, insofar as Descartes claimed precisely the dissimilarity between image and visible thing. Our visual perceptions are not the mirror reflections of nature, but projections of nature—this is the conclusion of Cartesian optics. The resemblance to which Berkeley refers, which constitutes the relation between map and visible, is not, therefore, the resemblance of two appearances, but the resemblance between appearance and essence.

What Berkeley is trying to discern in Cartesian optics is the notion that the image is not the image of the thing, but the effect of a projection that goes from the appearance of the thing to its essence, from the thing to its truth. As the projection of the visible, the image-map "catches" its "non-sensible" qualities and all the relations and proportions that form its essence hidden by its visibility: the image is the image of what is invisible within the visible. That is why the image "announces" a relation of nonsimilarity between itself and the appearance of the visible and a relation of similarity between itself and the essence of the visible. The image is, therefore, a good image, a copy.[7] The copy is never constituted by a similarity between the appearance of the copied and the copy. That kind of similarity is accidental and occurs among non-essential traits of things that could be essentially different. The copy can be maintained as copy only if it manages to copy the essence of the copied or its "idea," only insofar as it subverts "external similarity" and introduces an "internal" similarity with the copied object. When Zeuxis painted the grapes that seduced the eyes of the birds, when Parrhasios painted the curtain that Zeuxis wanted to remove in order to see what was behind it, their gaze was not deluded by the perfect similarity between the painted object and the painting, but by the fact that both grapes and curtain were "essential" copies: copies of the essence of the grapes or the veil.

However, the possibility that the bird's eye, which in the story of Zeuxis represents the natural attitude of visual perception, could be deluded affirms the curious fact that the logic of the "essential copy" is based on a "primitive" attitude of perception according to which continuous changes in the visible are the effect of its non-essential visible manifestation that is itself the effect of its non-changeable, stable, and hidden truth. This natural attitude assumes that all changes in the appearance of the visible, all histories of its surface, are merely the consequence of the ahistoricity of its depth that is the eternal geography of its truth. The paradox of the natural attitude of perception, therefore, lies in the fact that it never takes the visibility of the visible for granted, as if the eye does not want to be lured by the ruses of the visible, as if it does not trust the visible. For that reason, the copy has to "go" beyond the appearance, has to catch the invisible landscapes buried within the interiority of the visible and make them visible by its own similarity with them. The copy is the perfect copy precisely because it resembles the essence, and not the appearance of the copied thing.

By making visible the invisible essence of the visible, the perfect copy manifests its own "good nature"—it always fights for the truth. It "acts" like a detective who discovers the untruthfulness of the visible traces deliberately left by the criminal at the scene of crime in order to delude the investigation. In this struggle, the perfect copy defeats all delusions and openheartedly tells the truth. That is why representation (in its nature as essential copy) has the privileged status it enjoys in Cartesian optics. But also, and one should "perceive" here another fundamental reason for the substitution of pictures by images, the image is "good" because by copying the essence, it by the same token establishes the necessary connection between itself and model, sign and referent, exteriority and interiority. If the copy copies the relations and proportions that are nothing other than the essence of the visible, then its relation to the visible has to be essential: the relation of resemblance between essential copy and copied thing suggests that the copy possesses the essence of the copied—it is the "secondary" possessor of that essence. The act of copying doubles the essence of the copied by a paradoxical operation of actual multiplication of the essence that nevertheless remains numerically the same.

The existence of the essential copy, therefore, is proof of the existence of the good universe in which all things are necessarily connected, in which it is possible to "project" the truth of a thing, and then, thanks to the causal relationship between all things, to go from one truth to the other until the whole world is revealed in its reasonable, universal truth. That is why the existence of the essential copy is also proof of the existence of the good God who does not deceive us and who made the world like a fairytale forest from which one can go out into the light of the bright day simply by following any trace, for they are all good, all connected by the happiness of necessary connections. Good images exist only in the world from which the evil demon and its monstrous tricks have been excluded.

But, of course, nothing can be that good. That is why one is obliged to face a little problem here: the good images, the essential copies that Cartesian optics is trying to produce, do not exist. The good image is an outcome of the joyous play of fantasy that imagines the image is the visible presence of an invisible essence. That is the whole point of Berkeley's criticism of Cartesian optics. But this criticism finds us dangerously close to an unheard-of catastrophe: if the good image is only a fantasy, then, also, the good world of necessary connections exists only in the imagination. It was enough to climb the Vesuvius volcano on April 17, 1717, in order to see with one's own eyes that there is no such thing as the order of good, necessary connections. That is precisely what Berkeley did: "With much difficulty I reached the top of Mount Vesuvius, in which I saw a vast aperture full of smoak, which hindered the seeing its depth and figure. I heard within that horrid gulf certain odd sounds, which seemed to proceed from the belly of the mountain; a sort of murmuring, sighing, throbbing, churning, dashing (as it were) of waves."[8]

This picture of Vesuvius should be taken as a metaphor for the picture that is the proper object of sight. Standing on top of the volcano, we hear muttering, murmuring, or sighing coming from an unarticulated belly of the "truth," but we see only the smoke, only the flat surface on which there are no figures: we see only the picture, for the picture of the smoke is not the image or the "projection" of the depth of the volcano, any more than it is the "projection" of the smoke. We do not see the smoke as a geometrical form. Instead, we see the smoke, as well as any picture, as the amorphousness of all possible forms. We see smoke as the motion of the black, gray, or white color whose intensity can vary. Pictures are made exclusively of "light, shade and colours."[9] The picture is a trembling surface without depth, form, or figure. It is always unformed, because light and colors do not have form. Pictures are only different levels of intensity of light or colors. They are not "images projected on the retina." They are not projections at all—pictures are not copies.

Berkeley's determination that pictures are the proper objects of sight should be taken literally: pictures are *objects*. They do not have a model outside themselves. They are not "secondary" possessors of the essence of the object. Here emerges not only the fundamental idea on which Berkeleian optics is based, but also the outline of the Berkeleian world and its order: we are no longer dealing with the order of representation, but rather with the order of presentation. The picture is not a veil that covers visibility. It does not represent anything. The picture is the immediate object of sight, the visibility of immediacy, or the visible itself.

Every visible object is made as a juncture of certain minimal sensible visible elements—*minima visibilia*—and exists only as a particular collage of those minimal elements, or qualities of color, light, and shadow. The picture is an assemblage of those pure visual sensations, and the whole visible world is the assemblage of assemblages, an unstable complex of sensations. Instead of geometrical projections engraving themselves on the eye, what is behind the engravings now emerges: pictures made of sensible points, of *minima sensibilia*.

The eye sees the world in its innocence. Instead of the mathematized and ordered world, the eye can now see the world in its irregular motion. However, the disappearance of the labor of mediation between the eye and the visible means that the eye is never distant from what it sees, for it always sees by being affected by the picture, which is the collection of sensations. The eye sees by *feeling* the visible. It finally becomes the *sense* organ that feels the world because it feels the affections. It is exposed to sensations. It has to suffer the picture—the picture becomes the passion of the eye. Between the visible and the eye there is an absolute intimacy, for whatever feels cannot be separated from what it feels.

That is the paradox of Berkeleian optics: if the whole world is only the texture of sensations that the eye senses, then that world has to be *within* what senses it in order to exist at all, for it exists only insofar as it is sensed. The eye that suffers the action of the visible becomes, therefore, the condition of possibility for what acts on it. By its own passion, it gives life to the object of its passion. What acts can exist only on condition that there exists something that suffers it, and what is more, it can exist only within what suffers it. "Proper objects of sight do not exist without the mind. Whence it clearly follows that the pictures painted on the bottom of the eye are not the pictures of the external object."[10] They are the external objects that the eye senses *within* itself. The external objects exist within an "interiority" that is sensitive to them.

That is why the picture should be understood neither in its symbolic function, as representation of the sensible for the senses, nor in its imaginative function, as image of a resemblance. The picture is neither the picture of an object nor the representation of its essence. There are no originals other than pictures. The picture is the original with its own essence. It is, without a doubt, a "mad" object (because at the same time it affects the eye and is the outcome of that affection), but it is no less "objective" simply because it is mad. Vision is, therefore, an absolutely "objective" perception that knows no mediation, deformation, or misinterpretation. It is the effect of an immediate application of a picture to an eye or, what comes to the same thing, of a body to a body. Pictures are bodies made of visible qualities, bodies that can only be seen. Smoke is such a body. It cannot be touched or heard. Smoke does not whisper, and it is neither hard nor soft. It is a body made of colors: a completely immaterial body.

The whole peculiarity of the Berkeleian world resides here: in this world, there are bodies, there is no doubt about it, but those bodies are not assemblages of material elements, for in this world there is no matter. There is nature, but nature has changed its nature. Nature is not made of dead matter that cannot perceive. Nature is not an extended "stupid thoughtless something," as Berkeley would say. On the contrary, nature is alive and sensitive. It is a motion of sensations and their collections—bodies. In the immaterial, absolutely sensitive world, bodies are not made of material particles, but are the unstable collections of sensations. And some of those collections are pictures, visible objects, bodies, ideas. Between picture and object there is no difference: the idea is a thing is a picture is a body because all of them are nothing other than bundles of sensations: "as it is impossible for me to see or feel anything without an actual sensation of that thing, so is it impossible for me to conceive in my thoughts any sensible thing or object distinct from the sensation or perception of it."[11] The object-sensation is the immaterial materiality of the sensitive world. The picture, therefore, annihilates the distinction between essence and representation. It exists as original in a world in which that word has no meaning, for in this world there are only objects without representations of objects: it is a world that knows no copies. The picture is the original that represents itself by itself: it is the picture without similitude. That is why the picture is a simulacrum.

A simulacrum is not a bad copy or a failure of copying. It is not a vain effort of the picture to be similar to what it pictures. There is an essential difference between a simulacrum and a copy. A simulacrum is based on its own originality. It is a unique picture: "the simulacrum is built upon a disparity or upon a difference. It internalizes a dissimilarity."[12] That is why it is dissimilarity itself. The simulacrum is the "real" object defined by its own essence. It is the internalization of its own difference, which externalizes itself by itself within a space of disconnectedness of all simulacra: a sensation, no matter how similar it may be to another sensation, is always absolutely different from it and necessarily never refers to another sensation.

Simulacra, therefore, do not form a world in which differences can be connected into unity by causal connections that would enable the clear and distinct cognition of the world in its wholeness. Granted, the connected world of copies is also made of differences, but in it, differences are subjected to the labor of unification that always works in the service of identity. However, the world of simulacra is made of irreducibly different differences that cannot be causally connected. That is why in contrast to the copy, which is the good image because it mirrors the essence, the simulacrum is always an evil picture.

It is evil because it negates the good Cartesian world that offered the possibility of revealing the hidden connections between maps, the possibility of detecting the truth. In the Cartesian world, the existence of the subject-detective was possible, the detective who withdraws himself from the sensible world (always dangerously permeated by false, evil, and unconnected traces, "bad paintings"), and who, after this reasonable withdrawal, looks only at the good images in his own mind and, thanks to his perfect method, reveals the criminal. Like every detective, the Cartesian subject gradually, by the negative labor of his doubt, reduces the whole world until there remains nothing and nobody except the criminal or the "excellent" painting. "By the strength of logic alone, he has reconstructed the universe, and in his proper place has set the villain of the piece."[13] That is why Sherlock Holmes is the best example of the Cartesian subject. The fact that Holmes sees traces where the stupid inspector cannot see a thing, the fact that he never sees traces in the place of the crime, but always in the place where nothing happened, where there is nothing to be seen, means only that he, being the Cartesian subject, looks only for what he has already found. He is guided by the clear and distinctive insights that enable him to see the invisible essence of the visible. He is guided by the a priori knowledge that knows of the necessary connections between essences, thus enabling him to uncover the concealed relation between the criminal and a particular brand of cigarettes. Holmes is the Lacanian "subject supposed to know," for he knows that all traces inevitably lead to their cause.

On the other hand, simulacra are evil pictures precisely because there is no causal relation between them. Everything unfolds as if every simulacrum is nothing other than a minute machine ("ideas are formed into machines")[14] skillfully composed of different parts, of *minima sensibilia*, working as if in absolute self-sufficiency. The simulacrum is its own process of production and produces its own meaning: "one idea not the cause of another—one power not the cause of another."[15] The simulacrum is neither the cause nor the effect of another simulacrum. It can only be a good or bad *occasion* for another simulacrum to emerge. That the simulacrum can be a good occasion for the emergence of another simulacrum means that it can give it strength, joy, or happiness, but even then there is no necessary connection between them, for the occasion is not the cause. The occasion is only a contingency that is not necessarily realized. "Those [physical causes] may more properly be Called occasions yet (to comply) we may call them Causes. but then we must mean Causes y^t do nothing."[16] We can keep the term, but we have to change its meaning. We have to introduce the cause that does not cause anything, that does not work, does not produce consequences, and is not included in what is going to happen.

Simulacra, therefore, are causes without effects: abnormal, demonic causes. They move irregularly, compose sensation-objects, and more or less painfully vanish along the trajectories of their motion. This is the Berkeleian world, which comprises unforeseeable occasions out of which, all of a sudden, the sensation-object emerges. Those unexpected objects are provisionally connected into a fragile chain of contingency that constantly breaks off, thanks to the frictions of its links, thanks to the spasms of the sensation-objects that, after being "expelled" from the chain, continue to float on the turbulent ocean of simulacra, connecting themselves to another chain, participating in a vertigo of the genesis of an object. The world of simulacra is the vertiginous world of contingency, a monstrous world in which everything is possible: an evil world.

The Berkeleian person lives in this demonic world, which is demonic not because simulacra lie, but because they do not reveal the truth of the world in its wholeness. Every picture is true. The simulacrum does not hide anything. It is precisely what we see, since we see it immediately. But those pictures do not form a causal chain. There is no such thing as the universal truth of the world. That is to say, on the one hand, we are fortunate because we always have only true traces, but on the other hand, we are unfortunate because we cannot detect who is the criminal. The Berkeleian person never knows anything in advance. He floats from one truth to another, reaching in the end only the ultimate lie.

The world of simulacra is the pure force of truth that ends up in complete dissemination. That is why it is monstrous. In this world there are no detectives. Here, Sherlock Holmes is nothing other than a silly fantasy, and for that reason, the Berkeleian subject is less Sherlock Holmes than Bond, James Bond, the secret agent who is always already detected. James Bond is always transparent. The "criminals" always know where he is, who he is, what he is about to do. He is always inevitably involved in a vertiginous series of events that he cannot foresee: he has to go through them, he has to *suffer* them. The agency of this agent is submitted to the contingency of occasions in such a way that even in the end he will not know the whole story, for nothing comes to an end. There is no end, no story, since right now, at this very moment, something is happening that is going to change the structure of the chain of events. Bond is faced with evil pictures that are evil because they cannot be the traces in the process of detection.

The logic of the Berkeleian world, therefore, reads as follows: Every simulacrum is true, there is no reason for skepticism, everything is just fine—we will never reach the truth of the world. That is why, like Bond, the Berkeleian person has to go through the events only in order to find out that he cannot retroactively assign meaning to them because they meant what they meant only in the moment of their occurrence and because their meaning vanishes if they are connected into the whole. The demonic character of the world of simulacra lies in this frightening insight: everything could have been different, nothing can be foretold, there is no reason why something is, and there is no reason why something is not. There is no "why."

Archipelago and Ocean

The simulacrum—the Berkeleian idea—cannot be the cause of another idea because it is always passive. It simply happens as a thing, for, as we have already suggested, the idea is a thing or body: "The supposition that things are distinct from ideas takes away all real Truth."[17] It is therefore clear that existing things that cannot cause anything have to be caused, but by a cause that is outside of the order of simulacra. That is to say: "if the simulacrum still has a model, it is another model, a model of the Other."[18] Ideas are effects of the labor of the Other. God causes all ideas, he is the one who announces his acts in "pictures," who expresses and writes himself in the proper objects of sight. Every picture is, therefore, a "word" in this visual writing of God. This is Berkeley's conclusion, one that produces the analysis that has to produce that very conclusion: "I shall therefore now begin with that conclusion, that *vision is the language of the Author of Nature.*"[19] Let us, therefore, start from the end.

Every picture is a word. But in the book written in this language, the relation between picture and word or sign is not a relation of reference. The word does not refer to a picture, for that would mean reintroducing representation. In God's language, the word is not the commentary on the picture. The word does not possess the sense of the picture and does not name it: a word cannot be the title of the picture. And vice versa: the picture is not the illustration of the text. Looking and reading are not separate. If we wanted to draw a precipitous conclusion from this, we could say that God's visual language has the structure of a calligram. Granted: in a calligram, the "visible form is excavated, furrowed by words that work at it from within," but a calligram can never overcome its representative function, can never annihilate the distance between seeing and saying. Within the space of the calligram, reading, when it happens, produces meanings that can disseminate the picture. That is the "most perfect trap of the calligram": in spite of appearances, the calligram does represent. It is not the picture.

> Despite appearances, in forming a bird, a flower, or rain, the calligram does not say: These things *are* a dove, a flower, a downpour. As soon as it begins to do so, to speak and convey meaning, the bird has already flown, the rain has evaporated. For whoever sees it, the calligram *does not say, cannot yet say*: This is a flower, this is a bird. It is still too much trapped within shape, too much subject to representation by resemblance, to formulate such a proposition.[20]

Within the space of the calligram, the visible is not legible.

On the other hand, the fact that in God's visual language the idea is the thing, or its own referent, does not mean that the words of this language are made according to the photographic "regime." Although there are some similarities between photography and the simulacrum, the pictures that God makes are not photographs. Like the simulacrum, which is the idea that does not represent an *ideatum*, but is its own *ideatum*, photography is also tautological insofar as it "always includes its own referent within itself." But in contrast to the simulacrum, "photographs always connote something different from what they show on the plane of *denotation*."[21]

There is always something invisible within the visibility of photography. The nature of photography, therefore, is not that of an evil picture, but of a bad copy. That is why photography occupies the very place from which skepticism takes its force: it supports doubt concerning the existence of a sign that would be its own referent, in contrast to the simulacrum, which is always completely visible and which does not know of the difference between connotation and denotation. The simulacrum presents itself. It is the affirmation of the presence of a "now," whereas photography is the affirmation of the presence of the past. The tautological character of the photograph, therefore, is paradoxical: photography is at the same time both its own referent and the affirmation of a gap between the visible and the invisible. Photography is a "contact with what has ceased to exist" (what Barthes called a "funereal enigma"), a contact with invisibility, whereas the simulacrum is the transparence of pure visibility.

By the same token, this transparency affirms that the logic of God's visual language is different from the logic of the rebus. In God's language, the thing cannot be revealed by what conceals it, it cannot be veiled by its unveiling. It is not a language whose "word can take the place of an object in reality."[22] In divine language, there is no shadow above or behind the picture that could hide something in the picture or substitute for it. Every idea is entirely perceived. There are no reminders.

God's visual language is divine precisely because it abandons all calligraphic, photographic, and rebuslike distances between sign and picture, word and thing, "to see" and "to read." In this language, all gaps between essence and appearance are negated, which is what occurs in Paul Klee's painting. The visual language of God is made in accordance with the logic of Klee's surfaces. If Klee's surfaces are, as often has been said, barbaric, then it is because they do not represent anything. They do not presuppose a distance between language and world, visible and invisible. Those surfaces are located in a prelinguistic, visual language in which everything that was invisible has become visible by an unheard-of excess of visibility that established a passage leading from invisibility to visibility, thus negating the order of essences that exists behind the picture. Klee "undertook to build a new space,"[23] a space of visibility behind which there is no invisibility. His colored surfaces are the things in themselves. They are their own referents. They are located in the paradoxical place of the intersection of two different tasks: the "symbolic" task of a linguist, to make things invisible, to cover things with visible signs, and the "imaginary" task of a painter, to make things visible, to force things to emerge through the order of language that covers them. Klee's surfaces are placed precisely at the spot at which God writes his divine writing, at the spot that negates the sovereignty of the principle of distinction between verbal signs and visual representation:

> What is essential is that verbal signs and visual representations are never given at once. An order always hierarchizes them. . . . This is the principle whose sovereignty Klee abolished by showing the juxtaposition of shapes and the syntax of signs in an uncertain, reversible, floating space (simultaneously page and canvas, plane and volume, notebook graph and ground survey, map and chronicle). Boats, houses, persons are at the same time recognizable figures and elements of writing. They are placed and travel upon roads or canals that are also lines to be read. Trees of the forest file over musical staves. The gaze encounters words as if they had strayed to the heart of things, words indicating the way to go and naming the landscape being crossed.[24]

This indication is not signification. It is a suggestion or an occasion for something else—for the emergence of another surface. The "order" of this language also does not recognize denotation insofar as the idea is the *ideatum*, an original written by God's hand that writes the text-picture in the same way that Klee does: by sliding over surfaces of sensations that are word-pictures.

Instead of the order of denotation, there now emerges the "order" of expression. God explains himself before the eyes of all people, says Berkeley, and he explains himself by visual ideas. However, as the expression of a "model" (God), which belongs to a different order that cannot be presented, the expressed thing, in a paradoxical way, cannot explain God, who is trying to explain himself through it. God is the infinite set of all expressions, but expressions are not connected, and we lose the sense of the wholeness of the text: "'*Nihil dat quod non habet*', or, the effect is contained in ye Cause is an axiom I do not Understand or believe to be true."[25] The divine visual book is written in a stammering, disconnected language, in an absolutely first language that does not have any grammatical model or pattern to support it. It is written in the language in which, before any rules, there is the production of words, the building of the "flash" of the world. This language is an effect of an archaic labor that digs the word-thing out of the "phenomenological" pit of the absolute set of all minimal sensible elements, elements that are minimal because they are not collections in themselves, because they cannot express what has expressed them, but that, by being accidentally connected, form a word-picture, the sensual point of sense: an expression. The expression expresses only itself. Within this expressive language, sense is not something that lies hidden in the depth of the sign. The sign cannot be separated from the idea of sight. That is why it loses its character as signification and why the picture loses its character as representation—the picture becomes the presentation of the *sense* expressed within it as its expression. The sense emerges on the surface. It is the expressed of the expression. Every idea is, therefore, an expression of the sense.

The differences comprising the visual book of the world cannot be reconciled. The reconciliation of differences is the effect of representation and denotation, for only they can subsume difference under the identity of the concept and make of it a "conceptual" difference. On the other hand, God does not himself resolve the differences that are the effects of his own action because he does not live in self-identity. Differences emerge as intensive effects of God's expression, but God does not emerge as the expressive force that would be identical to itself. Granted, if God is the force of the production of the differences among sensations (ideas), then that force has to have an identity, but it is not an identity that would be the condition of possibility for the production of differences, as reflexive thinking would have it. On the contrary, it is the force that can be identified as difference in itself. God is the primal state of differences. He is the "differential" of every identity. The world or visual language, therefore, is a differential system in terms of which the distance between sign and picture vanishes, thus producing the simulacrum: "these differential systems with their disparate and resonating series, their dark precursor and forced movements, are what we call simulacra."[26]

Only within those differential systems in which identity is not split within itself by differences, but is the open set of differences, can the thing-idea be the solitary picture-word that expresses and makes visible only its own sense. As the expressed of God's expression, sensation presents only itself and its own sense: "What is expressed has no resemblance whatsoever to the expression."[27] We should insist on one thing here: Berkeley's concept of "suggestion"—the fact that the connection of ideas suggests the relation between the sign and the signified[28]—is not a reference to the regime of representation. The concept of suggestion does not mean that the sense of an idea is affirmed by another idea that takes the first idea for its object. On the contrary, the relation between sign and signifier is a discontinuous "connection," and in that sense it is not a relation at all. It is always a "relation" formed by the occasion, and not by denotation or representation. Signs "have no similitude or necessary connexion with the things signified."[29] The sign is the object in itself, and that is why it loses its "nature" as sign and becomes the simulacrum: it presents itself and is not necessarily connected to anything else, although it could be connected to everything else. However, there is no way to predict those connections, for there is no continuity: "things have no identity, if identity means durational continuity."[30] On the other hand, if identity means discontinuity, then things have the identity of the disparity distributed by the first differential—God. Every picture-word is accidentally mixed with another expression, and the infinity of those accidents makes up the vast texture of the text of the world. The world becomes a gigantic "*culinary fire*"[31] stirred up by God the cook.

God's visual language thus reveals itself as a schizophrenic language. Only the one who speaks that language is mad enough to claim that words are things, that things are ideas, that ideas are feelings, that feelings are things, that things are sensations, and that sensations are pictures, and so on and so forth. It is the only language that does not recognize the difference between bodies and words, nature and language. It is the only language in which the difference between nature and artificiality is not established by an "intelligible" border of sense. And since this language does not recognize representations, but only sensations, neither does it recognize specters and hallucinations. Hallucination or spectrality exists only within the gap between representation and its object, and not within the pure immediacy of presentation. That is why only a schizophrenic of the Berkeleian type can say: "There are no ghosts in the paintings of van Gogh, no visions, no hallucinations. This is the torrid truth of the sun at two o'clock in the afternoon."[32] It is the truth that shines or sparkles, that could be a torrid or ice-cold truth, but in any case the truth that can only be felt.

That is why simulacra are always true: if we feel pain, then it really hurts, if we see the white hotel, then it is the white hotel. We always feel what we feel that we feel. In this schizophrenic language, there are, therefore, only affections, sensations, and feelings that are good or bad occasions for another affection or feeling to occur: for an eye that stares into the night forced to rely on the frankness of touch, the night could be a pure and frightening darkness. However, for another eye that is looking at a body under the street lamp, the night could be pure white transparence. And so there it is—we have two nights at the same time, a glaringly white night and a grimly dark night. Each eye in this world has its own night, and in every eye there is an undeniable truth of a feeling. Truth and delusion are no longer competing, for there is no longer any representation: "In this primary order of schizophrenia, the only duality left is that between the actions and the passions of the body."[33]

That is why to say that there are no bodies in Berkeley's philosophy is a supreme nonsense. It is true that in this world, there is no matter, nothing insensitive. Which is only another way of saying that in this world there is nothing but bodies, for what is not a bundle of sensations, what is not a body, cannot be perceived and therefore does not exist. In other words, this world is completely "reabsorbed into the gap" that separated words from bodies. "There is no longer anything to prevent propositions from falling back onto bodies and from mingling their sonorous elements with the body's olfactory, gustatory, or digestive affects."[34] Words smell, odors talk, their talk is colored. Each color is a word, and each word has its taste. There are bitter and sweet words. One cannot differentiate anymore between two operational models: that of the body ("to eat and to be eaten") and that of the language (the existence of incorporeal events). Lewis Carroll's dilemma, "to eat or to speak," is now happily resolved—it is possible to eat words:

> But after all, say you, it sounds very harsh to say we eat and drink ideas, and are clothed with ideas. I acknowledge it does so—the word *idea* not being used in common discourse to signify the several combinations of sensible qualities which are called *things*; and it is certain that any expression which varies from the familiar use of language will seem harsh and ridiculous. But this doth not concern the truth of the proposition, which in other words is no more than to say, we are fed and clothed with those things which we perceive immediately by our senses.[35]

In common discourse, which maintains the difference between word and body, the idea is different from its *ideatum*. But this is not common parlance, this is divine language in which no word represents an *ideatum*, and no matter how harsh or ridiculous it might sound, it is nevertheless true that we are fed and clothed with ideas that are sensations that are bodies. Bodies are not concealed by language, but emerge on its surface: "in this case, we raise the operation of bodies up to the surface of language. We bring bodies to the surface, as we deprive them of their former depth."[36] Bodies become "surface" bodies, and words become bodies that present and affirm their own sense.

However, where there is no necessary connection between the simulacra, there is also no universal grammar, no universally valid law. God distributes words-bodies in series that are potentially contradictory within and among themselves, throwing them into a "mad structure" that we will call "the structure of the archipelago." The archipelago is a "mad structure" insofar as it is not structured by universally valid law. It is a changeable network established by "local laws" that connect minimal sensations into a collection or connect collections in bigger ones (smaller or bigger islands). Such "local laws" (every island can have its own law/s) are also unstable, always exposed to the force of still other laws that could disseminate the existing collection and "direct" its elements towards other collections. The archipelago, therefore, is made of fragile palisades "built" around the set of sensations that constitutes the set as a visible idea. Or, more precisely, the archipelago is at the same time the "formed" idea, the territory or the island, and the unarticulated ocean.

Archipelago: islands *and* water, the islands connected by water, which means separated by water, for water separates what it connects, and what is more, water does not respect firmly established paths. Water is the possibility of all possible directions, from any "drop" of water, one can go to any other point, to any other island, or else stay in the water, or the water can flood the island and destroy it. That is why the archipelago structure in every moment of its structuring suffers the labor of its own ruination: the green landscape of an island is shadowed and becomes a dark night landscape, or the intensity of light is stronger, and instead of the green landscape there is a red landscape heated to incandescence, the green gone red and torrid. But the green, red, or black landscapes are not different appearances of one and the same island. On the contrary, they are completely different landscapes, three different things, three ideas, three collections of sensible qualities, each of which is established by its own minor local law that will be abandoned whenever another law is put into effect.

The law of a collection is local both in terms of space and in terms of time. It is valid only within one collection and only as long as the collection exists. But since the collection is subject to constant change, it follows that the local law "rules" only here and now. The whole gigantic archipelago that is the universe is organized as the proliferation of an infinite number of simultaneous and different laws. Within the structure of the archipelago, it is always a question of how to turn a more or less complicated idea into something else, how to make of a milieu composed of different collections of ideas—for example of a bird, a tree, and a soil—another milieu composed of other collections. It is always a matter of incessant decoding and deciphering of a milieu that is nothing other than the production of another code. Every simulacrum produced by God is a cipher that deciphers a milieu within which it emerges, thus producing the occasion for a new milieu: the appearance of another bird within the milieu made of a bird, tree, and soil relies on decoding that milieu and establishing another made of two birds, tree, and soil. Or, vice versa, a milieu can be decoded by the disappearance of the simulacrum that was inscribed in it: the bird can fly away, and nothing will be the same anymore. Every milieu embraces an element that is just about to ruin it and to establish another milieu. The milieu itself is the outcome of a contingent encounter between its elements. The law of causality does not rule over it. That is why milieus are the simultaneity of differences, why they are not unitary: "The notion of the milieu is not unitary: not only does the living thing continually pass from one milieu to another, but the milieus pass into one another."[37] Everything can be transferred into everything else. A complex idea made of many complex ideas can be "transferred" into other "complex" ideas, thus making it even more complex. The whole universe comprises this orgy of promiscuous differences.

The only thing capable of determining the longer or shorter life of a milieu and thus preventing the complete dissemination of the universe into an amorphous ocean of *minima sensibilia* is rhythm. Rhythm is the duration of the interruption of distribution. It is the duration of *bonaccia*, the time when water is so still that it does not flood the island. Rhythm is the pause between two waves, the interval in the succession of two milieus. Like any other book, the book of nature has its rhythm, which is the rhythm of words and sentences, the rhythm of the succession of different milieus—the succession of ocean and land, city and desert, plants and animals: "there is rhythm whenever there is a transcoded passage from one milieu to another."[38] Rhythms, therefore, are the blank space between words or lines. Everything unfolds as if God has to think over what to write next in his work in progress, in his infinite novel that we call the world. Granted: every plot and every word is possible; there is no such thing as an impossible sentence. The only thing that is impossible is foretelling what will happen after the interval, for after the interval, anything can happen—all kinds of words can be said, thus forming different sorts of milieus.

Let us consider an acoustic milieu: somebody or something knocks on the door, "Knock, knock, knock." After that milieu, there is a pause, and the passage into another acoustic milieu occurs: we ask "Who's there?" Again a horrifying interval follows, an interval that is the occasion for any other milieu. It could be a good occasion, such as the appearance of a friend or a cute little kitten, or it could be a bad occasion, such as the appearance of an absolute catastrophe: "Knock knock: Who's there? Cat. Cat who? Catastrophe."[39] Rhythm is, therefore, precisely this: an endless abyss in which the ocean verges on touching an island, the possibility of all forms. From that "empty place" something will emerge to transcode the already existing collections: bodies, things, ideas, affairs. As interval, the rhythm is a diabolic moment because it is both the absence of forms and the possibility of all forms. The divine book is diabolic.

That means that the divine book does not exist as God's work. It is a network of forms connected in accordance with a law that unites it into the work. The work knows no interval, for the interval exists here, within the frame of the work, as milieu. The work always announces its end in advance, even when that end is surprising, insofar as the end narrato-logically follows from what has happened. The work obeys the principles of noncontradiction, sufficient reason, and identity that are going to mediate all differences. However, the Berkeleian universe is not a work, but the masterwork: "The work is made of forms, the masterwork is a formless fount of forms."[40] The masterwork is an archipelago, a text that can be read in all possible ways because nothing instructs us how to read it. In contrast to the work that is always good, the masterwork is always evil because it is the possibility of the simultaneous existence of all principles, rules, and laws, of all pictures, plots, and stories that are not progressing toward their more or less happy ends. And what is more, the masterwork has no knowledge of such a thing as an "end." It is infinite and always restless.

Rooms and Doors

Starting from the conclusion that "vision is the language of the author of nature," in the end, Berkeley arrives at the beginning: the visible is God's language. But how did we arrive at this "end—beginning," how do we know that God is the cause of the visible? The answer is short: we know that God is the cause of ideas because we know that we are not their cause, and we know that they can neither cause themselves nor one another: "For if I am conscious that I do not cause them, and know that they are not the cause of themselves, both which points seem very clear, it plainly follows that there must be some other third cause *distinct from me and them.*"[41] What seems very clear *is* clear, for there is nothing underneath the visible: the essence is the appearance. And if neither the eye nor the visible is the cause of the visible, then something else has to exist to cause the existence of the visible world. This is how we know of the existence of God. The whole proof of God's existence is based on this argument of "clear vision" and reads as follows: We clearly see that the visible exists. Therefore, God exists. And that is all as far as his existence is concerned. But what the nature of God is, whether he is corporeal or incorporeal, a thinking or unthinking being, one or multiple, we cannot tell for sure, because the nature of a God who is the cause that acts remains hidden.[42] His nature would be revealed if we could claim that there were a necessary connection between him and the visible world, but there is no reason to support that claim: "as to what you advance, that our ideas have a *necessary* connexion with such cause, it seems to me *gratis dictum*: no reason is produced for this assertion."[43]

On the other hand, the fact that there is no necessary connection between God and the visible world does not mean that there is a connection of similitude between them, for there is no reason to support that claim, either: "The cause of these ideas, or the power producing them, is not the object of sense, not being itself perceived, but only inferred by reason from its effect. . . . But we may not therefore infer that our ideas are like unto this Power, cause, or active Being." On the contrary, the only thing we may infer is that our ideas are completely unlike this Being, for "it seems evident . . . that in our ideas or immediate objects of sense, there is nothing of power, causality, or agency included."[44] Not only is it evident that there is no necessary connection and no relation of resemblance between cause and effect, but it is also obvious that in the effect there is nothing of the cause. It is reasonable to claim that between cause and consequence there is no causal relationship and that we cannot apprehend the cause from "its" effect. The cause remains hidden.

The visible is the expression of God and the means by which he explicates himself, but this explication does not explicate him. The visible is the proof of God's existence, but not the explanation of his nature. Or, to put it differently, the idea explicates only itself and by its mere existence implicates the existence of God. The connection between God and the idea, therefore, is as contingent as the connection between signifier and signified in the artificial languages of human beings. Of course, the difference between divine and human language could be defined as the difference between natural and artificial language, but that does not mean that natural language is less arbitrary than artificial language simply because nature speaks it. On the contrary:

> A great number of arbitrary signs, various and apposite, do constitute a language. If such arbitrary connexion be instituted by men, it is an artificial language; if by the Author of Nature, it is a natural language. . . . A connexion established by the Author of Nature, in the ordinary course of things, may surely

be called natural; as that made by men will be named artificial. And yet this doth not hinder but the one may be as arbitrary as the other.[45]

The relation between bodies is arbitrary, as is their relation with God, who is the "signifying being." Visual language (the world, nature) is the effect of the connection that connects only provisionally, the outcome of the cause whose principle of change changes constantly, of the cause that cannot be apprehended, determined, conceived. That is, however, precisely the proper "nature" of a cause: the cause is something that cannot be determined or known. It is constitutively indeterminable. Its indeterminability is the outcome of the way it acts, for it produces effects, but it is not what is determinate in the chain of effects, because the cause is not the law—the law in the sense of the natural law, that is to say, in the sense of the symbolic law. The cause acts by not acting as the law.

What emerges here is the fundamental difference between cause and law: the law is what is determinate in the chain of consequences (a collection of sensations, a milieu) because it belongs to the same order. It is within visual language. It determines what we call reality because it is itself within reality.

> Cause is to be distinguished from that which is determinate in a chain, in other words the *law*. By way of example, think of what is pictured in the law of action and reaction. There is here, one might say, a single principle. One does not go without the other. The mass of a body that is crushed on the ground is not the cause of that which it receives in return for its vital force—its mass is integrated in this force that comes back to it in order to dissolve its coherence by a return effect. There is no gap here.[46]

The law is integrated into what it makes lawful. The disintegration of the idea determined by the law is the disintegration of the law itself. From the determined chain of consequences a straight path leads only to the law.

On the other hand, the cause is not what is determinate in the chain itself. Or more precisely, it "emerges" in that chain as an "empty place," as what cannot be determined. It is unreachable from the order of effects because they do not belong to the same order. Between cause and effect there is always a gap that makes of reality—composed of effects—a limping reality, a reality that does not work properly: "Whenever we speak of cause, on the other hand, there is always something anti-conceptual, something indefinite. . . . Miasmas are the cause of fever—that doesn't mean anything either, there is a hole, and something that oscillates in the interval. In short, there is cause only in something that doesn't work."[47] Between cause and effect there is always an interval, a diabolical moment, the black hole of all possibilities, that makes of their relationship an arbitrary one. However, that their relationship is arbitrary means that access to the cause is foreclosed on the basis of the order of effects. That is why the cause can be determined only as something that is indefinite, only as *something*—as the real. The real is what cannot be assimilated into reality. The encounter with it always fails. Far from being the force that can determine reality, the real is what interrupts it. It is the discontinuity of reality, the annihilation of the law. The real is the trauma of reality. Only God is in the place of the real, in the place of the cause: "*the gods belong to the field of the real.*"[48] God is the trauma of the world.

So when he claims both the existence of the natural law and the discontinuity of nature, Berkeley does not claim anything contradictory. These two claims refer to two different orders, to the order of reality and to the order of the real, to the order of the law, in which there are no gaps, and to the order of "causality," which is made of holes, oscillations, and intervals that disrupt the order of natural laws and maintain the discontinuity of nature. As the explication that always fails, reality implicates the real as something that disrupts it, that emerges unpredictably in it, thus completely changing the connections within the visual field. And only from here can we grasp the real reason for the indeterminability of the cause. The cause cannot be determined because the cause is an accident. It is what hesitates and then acts abruptly and unpredictably. Of course, "the hesitation is not a matter of uncertainty but of 'illegality,'"[49] a matter of action that is not only hidden, but also not determined by any law. God's actions attack the law, thus completely restructuring its field. God is the accident that disturbs the law.

This means, in other words, that God, as what is indefinite and obscure, causes ideas that are definite and transparent, but that despite their transparency cannot show what they hide. We confront here the fundamental complication of Berkeley's philosophy. Where absolute certainty should rule, where there should be no room for skepticism, that is, within the "space" of the immediate objects of sight, an uncertainty now emerges. The absolute truthfulness of reality cannot convince us of its own truthfulness because the immediate transparency of truth is the very means of its own obscurity. By their absolute truthfulness, ideas can only show that their way to the cause is foreclosed, that beyond, within, or "around" them there is the real we cannot see, the truth we cannot catch. This, however, is not a failure of visual language. Ideas are successfully expressed by God, who does not express and explicate himself through what he expresses. God's visual language, therefore, is based on the failure of his self-signification. God, who constantly explains himself to the eyes of all people by "the arbitrary use of sensible signs,"[50] is therefore the God who "appears to himself," as it were, in his nature as cause, in his own indefiniteness—he is a traumatized God. He is his own trauma, because the truth of difference is the truth of God qua cause: the cause is the accident or chance, and chance implies the necessary failure of universal meanings. On the other hand, this "failure" is what induces God to endless expression. Thanks to it, God produces the world time and again, thus trying to explicate himself. God's incapacity to grasp himself is, therefore, the condition of possibility for God, for the world, and for visual language: this powerlessness is his power. If God were to become one with himself, he would no longer express himself. He would simply die from his own truth, and the world would vanish in an absolute disaster. Weakness is therefore the condition of possibility for the masterwork.

This whole maneuver that differentiates two orders was an effect of Berkeley's effort to restructure optical space. To put it simply: he tried to show that optical space cannot be based on principles of optics, but rather must be based on principles of semiotics. Only from the "perspective" of semiotically organized optical space (the cause that produces visual language) can the subject of vision legitimately ask the question what there is to produce the visible as absolutely visible while remaining at the same time withdrawn from that visible: "For beyond everything that is displayed to the subject, the question is asked, 'What is being concealed from me? What in this graphic space does not show, does not stop *not* writing itself?'"[51] What does not stop not writing itself is precisely what does not stop writing itself. Such is the insight of the semiotic reading of the visual field. Only semiotics can account for the invisibility of the hand of the author, for the absolutely arbitrary relationship between his hand and his text, for the fact that the Author behaves as a particular absolute: God "personally" writes his text, but this text should not be taken personally—in his own text, the speaking subject is not and cannot be the subject of speech. That is why his manuscript, even though it is called "the natural language," is an artificial language, a contrivance, a limping machine, and God is only an X, a mark of our nonknowledge: "Such is the artificial contrivance of this mighty Machine of nature that, whilst its motions and various phenomena strike on our senses, the hand which actuates the whole is itself unperceivable to men of flesh and blood."[52] However, if visual language is a machine that strikes (that acts), then this "striking" language has to be constantly changeable: a language without substantives, names, universals, or immobile bodies. The name presupposes a more or less frozen collection. It promises the possibility of classification. Names make of a liquid and mixed reality a succession of frozen "photos." That is why there are no names in a language in which all picture-words are the "strikes" or blows of sensations, in which all pictures are *verbs*. The verb-body that strikes the eye neither expresses another thing[53] nor represents an act: it is the act itself activated by God. God's visual language, that wild discourse, does not cover the world with the veil of words from human language.[54] It does not freeze bodies into an immobile furniture of the world. It is the language in which every picture (body, thing, idea) negates the distance between the eye and itself by striking the eye directly. Those "strikes-verbs" come from an *off*, from an invisible place that is the place of the cause. As if God were writing from inside his always already lonely room.

And it is only now that we can try to understand the way this God hides himself. Not all Gods are the same and not all Gods live in the same rooms. That is to say that the way they "hide" themselves is not the same. The essential difference between the Cartesian and Berkeleian God, for example, is that the former lives in a locked room, whereas the latter lives in a lonely room. The locked room is possible only in the world of causal connections in which depth has a necessary relation to the surface. Of course, as Cartesian optics has shown, this world is also based on the invisibility of depth, which organizes the visibility of the surface. But this invisible depth can become visible. This is the paradox of the locked room: even though the gaze sees and controls the whole room, the whole visual field, and sees everything that is in it, it is nevertheless possible to "deduce" something from that space that the gaze has not seen. So where does this "surplus of visibility" come from? How it is possible that when we unlock the room, we see something that was not in it when we locked it? The paradox of the locked room is resolved by the insight that depth does not have to be behind the surface, that it can hide itself on the surface, where it functions as the blind spot of the gaze, which organizes the whole visual field. What escapes the gaze is precisely that *the very room escapes it* because the room redirects the gaze toward the things in the room. What the gaze does not see is that not only are the things in the room, but that the room is within the room. This way of hiding depth is what makes of the surface a deep surface, the surface of perspective. This surface, therefore, always has an invisible "excessive element" (the wall of the room, God), which, by framing the surface, constitutes the blind spot of the gaze. Cartesian philosophy dreamed of the possibility of inscribing this frame into the interiority of the framed, of the possibility of knowing what escapes knowledge.

On the other hand, the lonely room is not locked. It is always open, but nobody can enter it. Or, the one who is in it cannot leave it, even though it is open. The lonely room is neither the depth behind the surface nor the depth on the surface. It is not connected with other places or spaces, and there are no corridors that lead to it. It is the depth that express itself in such a way that it is not present in what it expresses. It emerges on the surface only as the "friction" or the "spasm" of the surface, only when we "register a resistance to or failure of meaning."[55] It emerges only when a collection is disseminated into pre-individual singularities, into *minima sensibilia*, and it emerges as that very dissemination. The surface, therefore, is the flow: "In effect, if we mean by *things* the sensible objects, these, it is evident, are always flowing,"[56] and this flow is the constant failure of meaning. The paradox of the lonely room is that its frame, its "excessive element," is always already inscribed in it, but in such a way that that frame expands the interiority of the room into infinity, thus making of it something indefinite, something inaccessible. The words of visual language arrive from God's lonely room like the arrival of a flow, disseminating all nouns, names, adjectives, and still lives. In the universe made by a God who lives in the lonely room, nothing is still. There is only restless life—God's visual language is the primary and primordial state of language made exclusively of verbs of becoming. And "when substantives and adjectives begin to dissolve, when the names of pause and rest are carried away by the verbs of pure becoming and slide into the language of events, all identity disappears from the self, the world, and God."[57] The "logic" of the lonely room announces that neither God nor the world has identity. The world is the flux of the fluctuation of sensations that constantly negate the already existing assemblages ("complex thoughts or ideas are onely an assemblage of simple ideas")[58] so that every assemblage is continually faced with the possibility of its "nervous breakdown," of its negation. In the very moment in which we think we have grasped the meaning of an assemblage, its minimal elements start moving into completely different directions, and the collection vanishes. Everything is constantly in mutation and leads to the "transmutation of elements each into other."[59] And this transmutation, for its part, can cause a complete evacuation[60] of elements into another assemblage. If there is a grammar of God's visual language, then this grammar has only five rules: friction, mutation, transmutation, nervous breakdown, and evacuation.

Thanks to this constant flow, the paradox of the lonely room reads as follows: We entered the room, but that entrance was the exit. By entering the room, we found ourselves outside the room. The door in the depth, in the "origin" of visual language, thus remains forever closed in its openness. The door of the lonely room of the Berkeleian God, therefore, functions in accordance with the logic of Marcel Duchamp's "hinge." Standing in front of the door, in an exteriority, we enter another "side," what is behind the door, but only in order to find out that we have entered another exteriority. As an effect of a strange revolution,[61] the interiority is always elsewhere. Sometimes we think that it is on this or that side of the door, and sometimes we think that it is in the place of the door, whereas it is in the place of the "hinge." It is the hinge. It is something that is always in the place where we are not, for we cannot be where we are not. We are always only where we are, wherever we are.

Turbulence and Animal

To say that meanings continuously collapse or that the world has the structure of an archipelago is not to say that the world is chaos, but that it is born out of chaos. Chaos is a nebulous set, the pure inarticulateness that precedes every distribution. According to the interpretation offered by Berkeley, this chaos, an unarticulated whiteness of desires and powers that circulate through the infinite disparity of God's mind, is nothing other than invisible ethereal fire, warm light that creates everything visible: "The element of aethereal fire or light seems to comprehend, in a mixed state, the seeds, the natural causes and forms of all sublunary things."[62] God, therefore, is a mixed state in which everything that is not yet expressed vibrates. Chaos is the murmur of unarticulated sublunary things, the primitive secret, the pile of undistinguished pictures or words. That is why God's primordial chaos is named "fire" or, also, an ocean that is not yet intersected by islands. Fire is nothing other than the figure of unarticulated multiplicities of possible pictures, and the ocean is nothing other than an amorphous set from which earth and water have yet to be articulated: "These particles, *blended* in one *common ocean*, should seem to conceal the distinct forms, but, parted and attracted by proper subjects disclose or produce them; as the particles of light, which, when separated, form distinct colours, being blended are lost in one uniform appearance."[63]

Differentiated in itself, difference produces the world, the assemblages of colors. Conversely, undifferentiated difference loses itself within itself as in a common mixture of differences, of the nonindividuated motion of life. And it is precisely this motion of the oceanic mixture that is going to cause the self-explication of difference. What pulsates in the oceanic aquarium mixture of disparities or in the chaotic ethereal flame of blended forms is the vital flame, the motion of life, or the motion of God. The vital flame is what makes of the chaotic aethereal or culinary fire a fire from which different assemblages are going to be developed. In other words, the vital flame is both the principle that maintains the culinary fire as the indiscernibility of differences and the principle of the actualization of differences: "vital flame survives culinary fire *in vacuo*."[64] It maintains the indistinctness of differences that precedes every space, time, and place as a kind of "topology" from which space, time, place, distance, measure, and form are going to emerge, and through differentiation, the principle of this emergence actualizes the differences. When chaos starts articulating itself through the motion of the vital flame, then the book of the world emerges. Then emerges the language of God, who will actualize the unarticulated "ingredients" into the pictures-words-sensations of visual language: "sensations are also accounted for by the vibrating motions of this aethereal medium"[65] of the vital flame, which will compose the infinite number of irreducibly different pictures. "And it is very remarkable that this same element, so fierce and destructive, should yet be so variously tempered and applied as to be . . . the salutary warmth, the genial, cherishing and vital flame of all living creatures."[66] It is very remarkable that the unarticulated mixture of differences can differentiate itself, for this possibility is the possibility of the existence of the world.

The world emerges when the vital flame begins to differentiate differences. But this world is neither an absolute disorder nor an order. It is a state between chaos and harmony—it is turbulence. That the world has the structure of an archipelago means that it is turbulent: "The turbulent state mixes or associates the one and the multiple, systematic gathering together and distribution,"[67] in such a way that the one remains multiple within itself—an assemblage. The turbulent state means that from the oceanic mixture emerges an ordered island, an idea, a picture, a "local unit" that is an aggregate of sensations, a point of order within disorder. However, it also means that every such order is only a provisional utterance that will be refuted by another utterance whose possible articulation pulsates in the depths of the chaotic ocean. Every word of God's visual language—bodies, organs, plants, stars, minerals, animals, healing vapors, beneficial resins, books, rooms, and towers—is a disorder temporarily structured into an order whose particles attract and repel one another. The world is "an aggregate of the *volatile parts* of all natural beings . . . small particles in a near and close situation strongly act upon each other, attracting, repelling, *vibrating.* . . . Nor is the microcosm less attracted thereby. Being pent up in the viscera, vessels and membranes of the body, by its salts, sulphurs, and elastic power, it engenders . . . spasms, hysteric disorders, and other maladies."[68]

The macrocosm, therefore, is an endless living fabric, a gigantic animal in which there vibrates the force of life, thus provoking spasms and frictions of natural beings and bringing each "aggregate" back to the hysteric disorder. This is the "regime" of visual language: not "on," but within the "background" of the unarticulated "mass or mixture of most heterogeneous things" there appears a vibrating connection, a pulsating order. This exquisite state of arranged disorder is an effect of "animal" chaos, of the God-Beast, of the macrocosm-animal—the "world is an animal"[69]—which devours forms, swallows words, and reabsorbs them in its animal stomach in order to animate new forms. God is the cause (the invisible fire, the ocean) because he is an animal that "keeps up the perpetual round of generations and corruptions, pregnant with forms which it constantly sends forth and resorbs."[70] God, therefore, is in a paradoxical place: he is the pit of all potencies, and at the same time, he constitutes pictures and creates the visible world. God is both the chaos and the turbulence.

This release, reabsorption, and annihilation of all bodies (not only visible, but also palpable, acoustic, etc.) in the mind of divine animal occurs according to the only principle that the animal or life can obey, according to the unpredictable "principle" of "aversion and appetite" or affection and feeling. The divine animal cannot be affected by an exteriority, since every exteriority is within the interiority of this infinity. That is why the animal is affected "from inside." Through the chaotic motion of differences, chaos produces different composites of desires and feelings. That is to say, it produces different bodies (different sounds, colors, odors, forms): "And although an animal containing all bodies within itself could not be touched or sensibly affected from without, yet it is plain they attributed to it an inward sense and feeling, as well as appetites and aversions."[71]

Every word of visual language is thus an "embodiment" of an aversion or an appetite and its motion. It is this motion of aversions and desires that temporarily composes the unity of the world. This "composedness" is the paradoxical unity of turbulence. Within the "logic" of turbulence, it cannot be logically asked whether there is only the one whose predicates are multiple, just as it cannot be logically asked whether these multiplicities are "uncomposed," whether the world is deprived of forms. The logic of turbulence is rather: both the One and the multitude. The oneness of a thing is a constellation of multiplicities. The simultaneity of differences thus demonstrates that the unity is only an "optical" effect of the distribution of multiplicities. That is why the figure of this "composedness" is an infinite animal, an animal composed of other animals, or, also, a city that within itself gathers the multiplicity of bodies whose motions alter the constellation that constitutes the city: "And in this sense, the world or universe may be considered either as one animal or one city."[72] Animal and city are synonymous figures: every animal is a city because it assembles the multitude of other animals. Every animal is always already a menagerie. And every city is a menagerie, because it assembles the multitude of living bestial bodies. Both animal and city are figures of turbulence, of disordered order, established by the "diffuse principle" of longings, aversions, and affections.

However, the motions of desires and aversions could cause a collision of desires and bring God to exaltation, to the orgiastic delirium whose effect would be an absolute breakdown of meanings, a demolition of all constellations and relations: the ruination of the city that brings all bodies back into the stream of unconnected *minima sensibilia*. God, therefore, "remains quiet in remote cavities, till perhaps an accidental spark, from the collision of one stone against another, kindles an exaltation that gives birth to an earthquake."[73] The effect of this exaltation is a dramatic turn, a complete alteration of the visible—the entire world cracks in an earthquake, disappears in exaltation, all utterances of visual language are dispersed, and parts of the world can no longer be connected into their previous relations. It becomes clear that not only visible and local constellations, but the whole world is subject to the experience of a nervous breakdown. Earthquakes are the nervous breakdowns of the world out of which a new world emerges. And nervous breakdowns are the outcomes of the divine animal's desire, which brings itself to exaltation.

> Why may we not suppose certain idiosyncrasies, sympathies, oppositions, in the solids, or fluids, or animal spirit of a human body, with regard to the fine insensible parts of minerals or vegetables, impregnated by rays of light of different properties, not depending on the different size, figure, number, solidity, or weight of those particles, nor on the general laws of motion, nor on the density or elasticity of a medium, but merely and altogether on the *good pleasure of the Creator*, in the original formation of things? From whence divers unaccountable and unforeseen motions may arise in the animal economy.[74]

Moreover, the animal economy itself becomes uneconomical once motions of sympathies and antipathies are unforeseeable, once they are the effect of an exaltation of God's pleasure.

Abstraction and Crystal

Berkeley's critique of abstract ideas now becomes clear. To some extent, his critique has the same motifs as Spinoza's: the abstract idea is the negation of the difference between ideas or between finite modes. In Berkeley, however, the irreducibility of differences between ideas is not, as in Spinoza, secured by an individual "essence" immanent to every finite mode, but by a particularity of the constellation of sensible elements. The constellation that is an assemblage has no essence immanent to itself. Assemblages are "non-substantial," and their truth is in the uniqueness of the "joint" they constitute. "All knowledge onely about ideas":[75] there is no truth outside ideas, which is to say that there is no truth outside a particular idea, because "ideas . . . are all particular."[76] On the other hand, the particularity of an idea does not mean that it cannot be general. On the contrary, particular ideas are general insofar as every particular idea is a "joint" of minimal elements: as a joint of differences, every idea is general, and as the unique constellation of these differences, every general idea is particular. That is why "here is to be noted that I do not deny absolutely there are general ideas, but only that there are any *abstract general ideas.*"[77] To summarize two or more ideas into an abstract idea means to perform a fundamental negation, since the "generality" understood as abstract generality does "not consis[t] in the absolute, *positive* nature or conception of anything,"[78] but in the negative nature of nothing, in the negation of the existent particular.

Abstraction is guided by the idea of claiming the existence of nothing, the existence of fiction. Abstract ideas are fictions, or they are the fixed ideas of a mind crushed by a need to impose an order upon the unpredictable turbulence. To reject abstractions therefore means to reject the negative nature of fictions, to avoid sinking into the madness of hallucination. "I know there is a mighty sect of Men will oppose me. But yet I may expect to be supported by those whose minds are not so far overgrown wth madness."[79] I still may expect that in the minds of those who are not mad, of those who *feel* sensations, there will be no abstract ideas, just as I may expect that in their language there will be no abstract names, either. I may expect that in their world there will exist only particular "names" for particular ideas. This means: if an artificial language has to be established at all, then it should be based on a "reasonable" principle—"no word to be used without an idea."[80] The language of humans has to be established on this strict principle—a word for an idea.

However, this principle does not mean that the artificial word is the truth of the idea. It does not mean that it is possible to establish an equivocity between an "ontological" category (idea) and an "epistemological" category (for example, the feature of the idea), such as would be obtained through the mediation of artificial language. The same principle—every particularity has its concept—is also valid in the Leibnizian world, but in Leibniz, this principle means precisely that from a multiplicity, the One can be deduced as the truth of the multiple. In Leibniz, it is possible to "establish . . . equivocity between 'to be an element of,' or 'belong to,' ontological categories, and 'to posses a property,' 'have a certain predicate,' categories of knowledge,"[81] but that is because Leibniz's monad is different in nature from Berkeley's assemblage. Berkeley's assemblage is "desubstantialized," whereas the monad is the substance itself, which is why the multiple in Leibniz is not the representation of the labor of the One, but the expression of its truth. By thinking the monad as the multiple that envelops and is enveloped by the One, Leibniz could escape representation and formulate the principle that every thing has its concept. But Berkeley cannot do what Leibniz can. The assemblage is not the closed "house" that envelops the One, but the open, changeable constellation of multiplicities.

Berkeley's formula—a word for an idea—formulates the only way in which artificial language can come near to God's language while remaining, however, infinitely distant from it. Even in this best of all cases (in which there would be a word for an idea), artificial words would not designate the truth of the words of God's language. They would not affirm difference, but would perform the work of negation. An "artificial word" cannot be the expression of the idea, for even if it is the "proper name," the word still has the function of representation. It places itself in the *place* of the thing, and in this struggle between thing and representation, the unmediated truth of the world disappears. What has to be "sacrificed" in this struggle is the innocence of God's visual language. Artificial language thus conceals and negates the disparity of constellations.

The strategy of grasping truth is therefore identical to the procedure of dusting the house. The dust has to be wiped from the furniture of the world: "the chief thing I do or pretend to do is onely to remove the mist or veil of words"[82] in order to find the pure heterogeneity of the immediate sensations under that mist, a shine that would present the truth that dust is not the truth of furniture. "If Men would lay aside words in thinking 'tis impossible they should ever mistake, save only in Matters of Fact,"[83] because they will hold to sensations that are the only facts of the world, the truth that is felt and always felt truly, since sensations are always felt the way they are felt, no matter how they are felt. The removal of the veil of artificial language would suspend the labor of negation performed by representation and announce the affirmation of differences produced by pure presentations.

This is not to say that the (self-)affirmation of the sensations of visual language has the form of the negation of negation (the negation of the negative force of representation). The pair affirmation-negation exists only within the order of representation, as an effect of artificial language: affirmation and negation are related, to use Berkeleian terms, only to the affirmation or negation of words, not of ideas, and the difference affirmed or negated by the artificial language of humans remains subject to the demands of representation. That is to say it acquires the form of a *reflexive* difference that is subject to the "identity of the concept, the opposition of predicates, the analogy of judgment and the resemblance of perception."[84]

However, this quadripartite character of reflexive difference is grounded in the privilege given to the resemblance of perception. Reflexive difference, which culminates in the identity of the concept, constitutes itself as the outcome of the resemblance of perception that is the basis of identity, opposition, and analogy only insofar as the identity of the concept is acquired through a mediation of the negation of difference within the similar, insofar as the analogy of judgment is obtained through the "distribution of concepts," and not of sensations (analogy itself thus appears as the "analogy of identity in judgment"), and, finally, insofar as the opposition of predicates is established on the condition of the identification of an abstract idea, of an "indefinite concept" such as, for example, genus. In the "regime" of representation, the identity of the concept is the outcome of the resemblance of perception, which, however, cannot exist where there is only the coexistence of different sensations in which "knowledge or certainty or perception of agreement of Ideas as to Identity & diversity & real existence Vanisheth of relation becometh meerly Nominal of Coexistence remaineth."[85] Needless to say, resemblance also cannot exist where there is only the coexistence of perceptions for resemblance is an effect of a comparison of differences that occurs from "one" place identical to itself, which means that, in accordance with the phantasmatic loop of reflexive difference, resemblance is a consequence of what it itself causes—of identity. Comparison is therefore always "abstract" insofar as it mediates immediacy as a judgment about the immediate: "A man cannot compare 2 things together without perceiving them each, ergo he cannot say any thing wch is not an idea is like or unlike an idea."[86]

Comparison of similarities unfolds as comparison of representations produced by the labor of the negative that always presupposes the identity of the concept—the genus—and that has already "performed" the hierarchization of differences: "Abstract Ideas only to be had amongst the Learned. The Vulgar never think they have any such, nor truly do they find any want of them. Genera & Species & abstract Ideas are terms unknown to them."[87] The vulgar know only the vulgar truth, the truth of a feeling, and they feel only the distribution of coexistent ideas, which are neither identical nor similar to one another, but irreducibly different.

However, when the dust of words is removed and when one enters directly into the affecting motion of passive sensations activated by God, then affirmation and negation cease to exist. After this cleaning, no distance between idea and *ideatum* remains, but only an immediacy of perception, the absolute certainty of sensation that exists "before" or independently of every affirmation and negation. "'Tis allowed by all there is no affirmation or negation & consequently no certainty."[88] But what everybody allows is wrong. Or, more precisely: everybody is right to the extent that in "simple perception" there is indeed neither affirmation nor negation. On the other hand, they are wrong, for they do not see that the fact that in simple perception there is neither affirmation nor negation does not mean that there is no certainty in it. "Real" certainty lies precisely in simple perception, by means of which the affect and what is affected fall into one. Certainty is always and only the certainty of an "I feel," never the certainty of an "I judge" or "I think." That is why what everybody agrees upon "seems wrong. Certainty real certainty is of sensible Ideas pro hic & nunc. I may be certain without affirmation or negation."[89] What I see at this moment is absolutely certain because it is certain that I see whatever I see and that I *feel* whatever I see. "I cannot err in matter of simple perception" for the simple reason that I do perceive whatever I perceive.

The truth changes its meaning. It becomes certainty, and certainty becomes the only truth. Certainly, the truth as *adequatio* vanishes, because there is no longer representation, no split between idea and *ideatum*, but by the same token, the truth as *veritas*, as universally valid "insight," disappears or becomes meaningless: "Qu: whether *Veritas* stands not for an Abstract idea?"[90] *Veritas*, therefore, could be a mere abstraction, a nickname for a fiction. Truth, for its part, is never an abstraction. It exists only as the certainty of a singular event of sensation, and hence only comprehended certainty is knowledge: "Certainly I cannot err in matter of simple perception. So far as we can in reasoning go without the help of signs there we have certain knowledge."[91] Everything that is covered with signs is in the region of uncertainty, and we attain knowledge only insofar as we hold on to the singularity of sensations and feelings: we know only what we feel.

The "truth" that we know only what we feel is an inevitable effect of the very nature of the "primordial difference," which is God. It is the consequence of the God-animal, who "replaces" the truth of the universal with the certainty of the singular because he is the power that emits the simultaneity of incompossible events. This is Berkeley's answer to Leibniz. Leibniz says that "it is not the impossible, but only the incompossible that proceeds from the possible,"[92] and that therefore in the same universe there cannot exist worlds or events that are not simultaneously compossible. However, if God is the primary difference that emits the simultaneity of incompossible events (turbulence), and that simultaneously "traverses" these incompossibilities, then nothing prevents us from claiming that what is not simultaneously compossible belongs to "the same" world. Thanks to the turbulent nature of God, we can claim the insane thesis: what is not simultaneously compossible can be simultaneously compossible. Such is the radical consequence of the insight that sensation is the only possible certainty and that truth has the form of this certainty.

Here Berkeley anticipates an "abstract" objection. If knowledge is nothing other than sensation, then Berkeley's writings, which are written in the artificial language, are not true, and his insight that there is no universal truth is a deceit. Sure, replies Berkeley, whenever I am using language, I am using untrue abstractions, but from there it does not follow that knowledge is not in the sensation. On the contrary, it follows only that sensation cannot be uttered. And that is why I "warn him [the reader] not to expect to find truth in my Book or any where but in his own Mind. w^t ever I see myself 'tis impossible I can paint it out in words,"[93] because "painting in words" is a representation that negates the "painted." Truth as certainty of the intensity of sensation lies only in a language that does not represent anything, in the visual language of God, which reveals the true nature of natural language, namely, that language, because it is turbulent, cannot have "firm" grammatical rules, that it must have the nature of a "flexion." Not re-flexion, but flexion. Change and affect produced by this change is the nature of God's language, and that is why we see nature in its entirety as a "set" of incompossible scenes in which all visible things "appear" as "fluent and *changing* without anything permanent in them."[94] One can no longer say that flexion is just a way in which language mimes bodies, which is the "function" of flexion in artificial human language. This "miming" is possible only in a space of representation where artificial language reflects flexion within itself, thus making of it a reflexive flexion, miming its nature "for" the mind, so that the mind can grasp the body "at a distance." By reflecting the flexion of the body, reflexive flexion annuls the flexion of the body: language itself becomes a bodiless body, and the sensuous body is expelled into the depths of silence. That was the strategy applied in the Cartesian mapping of the world. But, as was already shown, the representation of the sensible through the language of maps, by means of which language tries to take into itself the nature of the body, led to a gradual superimposition of grammar upon flexion that finally reduced flexion to a mere grammatical principle. The flexion of the body became a stable geometrical-grammatical figure, bodies were expelled into the "petty" silence of the "flesh," and language became the pure/puritan language of the "beautiful soul" searching for universal truth.

But now, as an effect of removing the veil of representation, there is no longer the "petty silence" of the speechless body and the "pure speech" of bodiless language. When words are bodies, then language no longer hides anything—God's visual language is a pure flexion of the body: bodies are talking now. Language is made of bodies. Of course, this language is essentially different from the language that represents it. It is correct that the language that represents bodies through reflexive flexion also uses "the figural" and its connection with the visual in order to establish itself within the chain of the figural and nonfigural. It is also correct that this connection is established through a relation between representation and represented such that the presented, say, a body, can change its representation, as was seen in Cartesian theory, on the basis of which it was possible to assert that "figures" or the signified are images that form language and that representation is the outcome of presentation. But the effect of connecting the figural and the nonfigural was such, as was also seen in Cartesian theory, that the nonfigural, brought into language by the figural, "overcame" the figural. The nonfigural reduces the figural to a kind of drapery thrown over the supporting wall of language. This wall, which secures the marginalization of the figural, is built not of paradigms, but of syntagms insofar as the syntagm is the condition of possibility for the homogeneous and "cumulative" nature of language, whose "structural order" functions as a kind of the causal bond between the elements of the syntagms and the syntagms themselves, thus producing the narratological logic of discourse, its function of identification, its work of connecting differences into one.

The transgression of language into the body that occurs in God's visual language implies, however, that the syntagm, which bound elements in an organic totality of narration, no longer organizes the structural order of language. God's visual language does not have a narrative structure, which is not to say that it is a senseless circulation of sensations: turbulence is neither chaos nor narration, but non-narration. Narration always appears as what seizes the truth insofar as its development presupposes the existence of both spatial and temporal relations between distributed elements (events, plots, stories). By connecting those elements, narration imposes itself as truthful narration: "Truthful narration is developed organically, according to legal connections in space and chronological relations in time. Of course, the elsewhere may be close to the here, and the former to the present; but this variability of places and moments does not call the relations and connections into question."[95] It does not call into question the homogeneity of the totality because the variability of places and moments is connected into a single point of view through their mutual relations. Or, to put it differently, this single point of view—God's apperceptive monad, for example—gazes at the totality of different relations and constitutes the plane of their truth. But in the non-narrative structure of God's visual language, narration is not promised at all. The simultaneity of heterogeneous stories will never be overcome through their homogenization into an organic totality of truthful narration because what structures the narration is missing—there are no relations. In the non-narrative structure, the relation is denounced as an indistinct term, as an abstract idea: "The obscure ambiguous term *Relation* w[ch] is said to be the largest field of Knowledge confounds us, deceives us,"[96] and all one has to do in order to "remove" this deceit is to "undress it,"[97] to look under the words and reveal the truth that there is no truthful narration and no organic totality—that there are no relations.

In God's visual language, everything unfolds as a kind of artistic practice in which different spectacles are not reduced to a single vantage point, but each constitutes for itself its own "fluctuating" center, thus releasing the field from vision of the difference between center and periphery. Or, everything unfolds as in a cinematic practice of distorting frames, of deframing, of radical displacement of the point from which one looks, which breaks the narrativity of "continuous framing" into a non-narration. Simply put: the non-narration of God's visual language means that within it, the stories multiply, but there is no narration.

The non-narrative nature of God's visual language is possible only because the picture is essentially different from the image as representation in the field of artificial language or in the field of geometrical optics. In artificial human language there are only "organic" images. The organic image is nothing other than the name that brings the nature of representation to light, revealing the fact that it is always a description of the object, which exists independently of it: "A description which assumes the independence of its object will be called 'organic.'"[98] In other words, representation will be called an "organic image" because it covers the object and thus makes it invisible. The organic image can be infinitely large or infinitely small, as Leibniz had shown, but regardless of the "size" of its infinity, it is always a closed, continuous totality: "In an organic description, the real that is assumed is recognizable by its continuity—even if it is interrupted—by the continuity shots which establish it and by the laws which determine successions, simultaneities and permanences: it is a regime of localizable relations, actual linkages, legal, causal and logical connections."[99]

None of this exists in the language that is the flexion of the body. In this regime, reality cannot be recognized by its continuity insofar as in this reality, identity is a discontinuous "connection" and the absence of permanency and succession: "Identity of Ideas may be taken in a Double sense either as including or excluding Identity of Circumstances, such as time, place etc."[100] What is announced by this insane utterance? Nothing less than an unheard-of madness: namely, that it is possible (and it is enough that something is possible for it to be actualized) that the identity of the picture can exclude the identity of time and space and also of all other circumstances, thus scattering the identical into differences, into discontinuous and unrelated places and moments. For this reason, the organic image disappears entirely. Reality can no longer be the outcome of organic description, and the circle in which the reality of the object is the effect of its representation, that is, the effect of reality of the object represented by it, is broken. There remain only pictures that are objects or bodies. There remain only fragile "crystals," insofar as the crystal is the transparent, visible multiplicity of indifferent differences.

The universe is a gigantic distribution of crystals. From such a distribution of crystals, which does not constitute an organic unity, no identity of the world can be established. This impossibility could be the basis for an interpretation of the fundamental principle of Berkeley's philosophy that would read as follows: In his interiority, God produces and perceives a visual language as an exteriority that is his own interiority, which exists only as long as his perception lasts. *Esse est percipii* would therefore mean that God produces the world as a perception to which nothing in exteriority "corresponds" because exteriority is precisely that perception. In other words, it would mean that God produces the world as his own hallucination. The world would exist only within the region of God's imaginary, in which "subjective" images act as objects and actual images cannot reach their own actualization.

Of course, nothing is more erroneous than this interpretation. The imaginary presupposes the existence of a "reality" that is "different" from it. The imaginary is what falls out of the harmonious order of reality—a deceit. But in the world made of crystals, the horror caused by the possibility that the world is a mere contrivance of divine hallucinating perception disappears—not because reality is lost in the imaginary, and still less because the imaginary is maintained as the only reality, but because God's visual language is the point at which "the imaginary space and the real space fuse,"[101] a paradoxical point at which real and imaginary become indiscernible, at which "subjective" images, while remaining "subjective," reach their own actualization, thus becoming at the same time "objective."

This is to say that God's visual language has the nature of a rainbow: "When you see a rainbow, you're seeing something completely subjective. You see it at a certain distance as if stitched on to the landscape. It isn't there. It is a subjective phenomenon. But nonetheless, thanks to a camera, you record it entirely objectively. So, what is it? We no longer have a clear idea, do we, which is the subjective, which is the objective."[102] Our clear idea of the difference between subjective and objective, virtual and actual, is undermined because visual language is the effect not only of God's production, but also of God's perception, and because all things in the world have features both of "virtual" and actual images. Everything in the world therefore has the paradoxical nature of an "image" that does not have characteristics of an object insofar as it is an effect of God's "subjective" perception, and of an "image" that has all the characteristics of an object, all the features of a "natural thing," as Berkeley puts it, insofar as it can be seen "directly," and this always means insofar as it can also be seen "indirectly," in a mirror that reflects a virtual image of an actuality.

Virtual and actual coincide, thus proving the Berkeleian-Beckettian thesis that "self-perception maintains in being" not as self-reflexive apprehension (that would mean that visual language is made solely of virtual images), but as a field in which God's eye (E in Beckett's screenplay) and the eye of his object (O) merge. It is a field in which the clash between E and O arises every time that O enters *percipi*. Quite simply: we do not have a clear idea about what is subjective and what is objective because God "organizes" his language in accordance with the crystal regime in which actual and virtual fall into one another: "it is here that we may speak the most precisely of crystal-image: the coalescence of an actual image and *its* virtual image, the indiscernibility of two distinct images."[103] It is here that we may speak of the coalescence of the gaze of E and "his" image (O), of the coalescence that enables God's eye (E), which has produced O from within itself, to observe O as an actual object of its perception, as virtuality that has fallen into one with its actuality:

> E advances last few yards . . . and halts directly in front of O. Long image of O, full-face . . . sleeping. E's gaze pierces the sleep, O starts awake, stares up at E. Patch over O's left eye now seen for the first time. . . . O half starts from chair, then stiffens, staring up at E. . . . Cut to E, of whom this very first image (face only, against ground of tattered wall). It is O's face (with patch) but with very different expression, impossible to describe, neither severity nor benignity, but rather acute *intentness*.

Of course, there are infinitely many objects of perception, because God is heterogeneous. He is an eye that has infinitely many pupils. He is the light, which constitutes all objects of vision and which is heterogeneous, too. Light is a "heterogeneous medium, consisting of particles endued with original distinct properties,"[104] a "medium" that forms ideas as heterogeneous collections and then gives them "back" to God's gaze so that it can see, simultaneously, the open totality of the world. This is why Berkeley can say that "different sides of the same ray shall one approach and the other recede from the Icelandic crystal."[105] This paradox, according to which one and the same ray at the same time approaches the crystal and recedes from it, explains the paradoxical nature of God's gaze and the paradoxical nature of "pictures" as actual-virtual objects. Produced by the very gaze of God and therefore deprived of distance from that gaze, pictures are nevertheless at a distance from it. They are distanced by that very gaze, which distances or actualizes them in order to see them. That is why they are actual-virtual. They are crystal pictures in the crystal that produces them, which

Berkeley calls the "Icelandic crystal." And God is that Icelandic crystal, the crystal of all crystals, which distributes crystal pictures. God is the crystal ball of difference: both chaos and order—turbulence.

Scenography and Iconography

However, the fact that God sees everything (and there is no doubt that he constantly perceives the world, for the "eyes of the Lord are in every place")[106] still does not tell us anything about the way he sees. If God is the viewer of his work, then surely this viewing cannot occur in accordance with geometrical optics, insofar as it explains the establishment of the visible by referring to the imaginary and the labor of representation. "For there to be an optics, for each given point in real space, there must be one point and one corresponding point only in another space, which is the imaginary space."[107] Optics presumes the existence of the imaginary, which does not exist in God's mind because his mind is a paradoxical crystal that makes pictures of both actual objects (distant from the eye) and virtual objects (within the eye). That is to say, the objects of God's gaze are at a distance from it, but this distance is only "one point in the fund of the eye."[108] It is a not-distanced distance.

On the other hand, only a distinction between real and imaginary space enables the object to be established as the object that is "really" distant, somewhere there, in another space, in the *depth* of that space. Only that distinction produces the depth of the perspective image as the depth within the image, and not the depth of the image. The image is always a flat surface, a two-dimensional plane represented for the eye as a three-dimensional image through the system of projections of optical laws. Of course, in order for optics to maintain this organization, it is necessary that, as was the case with Descartes's gland or cerebral cortex, the flat surface of the image be posited as a "plane" that at its center connects different "planes" under a certain angle, that connects the "diagonals" of these "planes," thus creating a background and a foreground, depth and surface of this depth. The depth of a flat image has the function of the depth of reality for the imaginary. It represents reality for some imaginary.

That is why the depth within the image functions according to the principles of scenography: the object of vision is an isolated scene offered to the gaze, a scenographically created fragment of the world. To the gaze is given a "depth of reality" posited in a fixed place within the absolute space of the world. Hence another presupposition necessary for the constitution of scenography emerges: the distant object of vision, the production of the depth of an image, does not require only the existence of the imaginary and the intersection of diagonals of the "planes" of the image at a central point. It also requires a clear distinction between the place and absolute space, an irreducible difference between relative and absolute space.

"To-day those who discuss motion . . . postulate space on all sides measureless, immovable, insensible,"[109] calling it the true space and distinguishing it from relative space, from the *locus*, from the space of the body that moves. They determine the *locus* only as phenomenal space. But since the eye experiences the perspectival image, and scenography shows that there is no absolute space, that there are only relative spaces, then the perspectival image is only a deceit of the eye. Absolute space cannot exist "because all its attributes are privative." They are "mere nothing," and therefore they are not perceivable by the senses.[110] The only "true" spaces are always relative places determined by the relative motion that occurs between two "correlated bodies."[111] That motion is "transferred" from one body to another, thus changing it by changing the direction of its motion. Relative place thus is always changeable, affected by relative motion that puts both correlative bodies into motion.

However, the motion that occurs between the eye and the perspectival image produces the effect of depth only on the condition that their relation remains unchanged. If perspective were to be based on relative motion, there would have to be a relative motion between an *immobile* image and the eye, which, according to the nature of relative motion, is impossible. A deception performed by the perspectival or scenographical image is disclosed here. When we move the gaze from the image, it does not move with the gaze, but remains immobile, and so manifests itself as the "untrue place," as Berkeley would say. For scenography to exist, it is necessary that the *locus* have the function of the place, but at the same time lose what makes it a place—mobility. Only on this condition does the visual field appear as scenography. Only on this condition can a "landscape" or an image appear out of the amorphous background of absolute space with a clear "edge" that corresponds to the "frame." This, at least, is how Leibniz explains the deceitfulness of scenography based on horizontal or central perspective.

Imagine a traveler who approaches a city from one direction or another, says Leibniz. What this traveler sees as a city, in the moment of his approaching, is an image organized "around" horizontal perspective: the city "delights the eyes of travelers . . . with horizontal perspective."[112] The city appears to the eye as a perspectival image or as a scene that the eye views as if it were at the top of a "cone." The eye looks at the scene in such a way that from where it is situated all the way to the center an axis is established—a "central" ray. The traveler sees it as a scenography. Of course, to travelers who are approaching the city from different directions, the city appears as different scenographies. Every perspective has "its own" scenography, and because they are only different scenes of the same, none of them is the "truth" of this same, none of them is the truth of the city. Scenography differs from the city in the way a distinct idea differs from a confused one. This is to say that scenography deceives the eye in the manner of appearance, and not in the manner of illusion, for it is neither a hallucination nor an object that could maintain its "truth" without being mediated by the finite gaze. It is not an object that God's eye sees in the innocence of its unchangeable truth. Conversely, it is constituted through a cooperation between the eye and exteriority and varies depending on the place of the eye—it is an aspect: the "external *aspect* of a city varies as you approach it differently, from the west or from the east, the qualities of a body vary with the variety of our sense organs."[113] That is why scenography deceives: it is a sensible quality, not, however, apprehended à la Berkeley as the object in its truth, but as a quality that is both the relation between the eye and the visible and a form of self-relation of the eye. It is for this reason that Leibniz calls scenography an "aspect." Aspect preserves this fundamental twofoldness: "*Aspectus* is at once gaze, sight *and* that which meets the eyes."[114] It is at once the spectator and the spectacle. The fact that in Latin, *spectaculum* is precisely "the gaze" bears on this self-reflexive folding of the aspect: in the form of the spectacle, the gaze wants to reappropriate itself, to see itself. Scenography exists only if the gaze acts against an artificially established center of the visible, against the vantage point with which it is connected by a "prince ray" in an "almost infinite horizontal perspective" in such a way that the "center" is precisely the point of the gaze, and by looking at it, the gaze looks at itself and mirrors itself, as in a kind of infinite "quasi" reflexivity. Scenography is the site of the untruth because it is always an aspect or a spectacle, a self-mirroring of the gaze, and not a mirroring of what is offered to the eye.

But let us imagine a city that has a high tower at its center. Let us imagine that a gaze is already in the city, at the top of the tower. This will give us the image of God's gaze. God's gaze is at the top of this tower, and the tower itself is a straight line that connects God's gaze and the center of the city, in which, according to Leibniz's assumptions, all "almost infinite horizontal perspectives" intersect. The place at which God's gaze is posited, therefore, is no "bird's-eye perspective," but an "infinite" perspective that vertically pierces all horizontal perspectives and gathers them into one. God's perspective, therefore, is a kind of an *internal perspective* of all the external aspects of the city, the place from which all scenographies are viewed from the "inside": from this place, the unchangeable essence of what deceitfully appears in scenographies is seen. If one looks at the city "from a tower placed in its midst . . . this is as if you intuit the essence itself."[115] God's gaze sees the essence as well as appearances, spectacles as well as the spectators. All scenographies and all gazes are now posited within the interiority of God's perspective, in a situation that inverts the operation of horizontal perspective. This "inversion" means that the world is not presented to God's gaze in such a way that that gaze would be withdrawn from it, as the eye is withdrawn from the visible within the logic of central perspective. Conversely, God's gaze is in the very center of the painting or of the world, like the tower in the center of the city. God's "internal perspective" is, therefore, an inverted perspective.

Finite monads are condemned to scenographies. They are condemned to be manifold expressions of the universe from a certain point of view. They are condemned to grasp clearly only a fragment of the world and only "unclearly to move towards the knowledge of the infinite." And even though the finite monad constantly changes, the point of its "view" is single and static in each moment. "A certain point of view" of the monad, therefore, is organized according to the principles of central perspective. Contrary to this, God "in regulating the whole has had regard to each part, and in particular to each Monad."[116] At the same time, God sees all points of view, which is precisely to say that his gaze is not determined by "a certain," static point of view, that his gaze is dynamic, that it is the motion of a multitude of gazes. This is the key difference between central and inverted perspective: "The system of inverted perspective results from the use of a multiplicity of visual positions, which is to say that it is connected with a dynamic visual gaze and a subsequent summation of the visual impression that is received in a multilateral visual embrace. . . . The opposition of the systems of direct and inverted perspective may thus be linked, above all, to whether the visual position is fixed or, conversely, dynamic."[117]

From this opposition there follows another: the dynamic of the gaze enables God to see "things" simultaneously from all sides, to see what they *are*, to see the very "concentrate of Being," in contrast to the finite monad, which, determined by its point of view, can represent infinity only indistinctly. Its "representation is merely confused as regards the variety of particular things [*le detail*] in the whole universe."[118] It does not seize infinity, but only its indistinct representation, which is the outcome of what appears in front of a finite gaze. This is, in fact, the crucial difference between central and inverted perspective: "It may be said that in the first case [inverted perspective] the essential thing is what the represented object *is*, while in the second case [central perspective] the essential factor is what it *seems to be*, that is, how it *appears* to the artist's eye."[119]

Or, to put it differently, the eyes of a traveler who approaches the city are focused on a part of the whole, whereas God's gaze views the wholeness.

This introduces another difference between central and inverted perspective: "Indeed, if a picture is painted according to the rules of direct perspective, an object represented in it can always be extracted from the picture," in contrast to the painting based on inverted perspective, where the "artist's task is, above all, to represent such a world *as a whole*."[120] This difference between the finite gaze and God's gaze, which we are trying to articulate as the difference between central and inverted perspective, means that the finite monad sees the world as scenography, whereas the infinite monad sees it as icon.

The icon here should not be understood in the sense of a painting, sign, or image that to a finite gaze announces the presence of God's gaze on the other side of it, beyond the representation. Rather, the icon here appears in its original meaning, as a window that is the very gaze of God. According to Pavel Florensky, this "window giving us this light is not itself 'like' the light, nor is it subjectively linked in our imagination with our ideas of light—but the window is that very light."[121] Or, to put it differently, the icon is not similar to the "heavenly vision" of God, but is this very vision: "an icon is the same as this kind of heavenly vision." However, in the regime of iconography described by Florensky, the possibility of inverted perspective is not offered to God's gaze alone, but also to a finite gaze, which can view the icon, whereas in Leibniz, God is the only one who views the infinite universe as if it were the icon: God is the one who looks at his own gaze—an infinite apperception.

It is precisely this self-reflexivity of God's gaze that distinguishes Leibniz's inverted perspective from iconography. In contrast to the Byzantine inverted perspective, where the finite gaze could enter the "interiority" of the icon, since the icon was "the visible image of . . . supernatural visions,"[122] Leibniz tried to make an iconography only *for* God. By introducing all the features of inverted perspective, but reserving them for the gaze of God, he did not, in the end, establish iconography, but a devalued central perspective with all the features of inverted perspective and a single feature of central perspective, precisely what makes it central perspective—the self-reflexivity of the gaze. By establishing the vertical axis of God's gaze in which all horizontal perspectives intersect, Leibniz tried to overcome scenography: "Leibniz collected scenographies of a given thing in order to seize its iconography,"[123] but it escaped him because he established it as a self-reflexive structure centered on a vertical axis that returns God's gaze to itself. "Order demands that curved lines and surfaces be treated as composed of straight lines and planes, and a ray is determined by the plane on which it falls, which is considered as forming the curved surface at that point."[124] Leibniz, therefore, assumed an order of flat planes and straight lines as a primary order—he assumed an iconography. And after that, he preferred to it a scenography:

> But the same order demands that the effect of the greatest ease be obtained in relation to the planes, at least those which serve as elements to other surfaces. ... This is all the more true since it thus satisfies, with respect to these curves, *another principle which now supersedes the preceding one*, and which holds that in the absence of a minimum it is necessary to hold to the *most determined*, which can be the *simplest* even when it is a *maximum*.[125]

In other words, the method of maximum and minimum, which Leibniz applied in his catoptrics, presupposes the shortest ray's path that falls on the plane, a hypothesis that also "works in the reflection itself when applied to curved surfaces and in concave mirrors." To put it simply: "universal truth" always turns out to be that the gaze is directed by "the most determined or unique path"[126] and that by "falling" on a flat plane, the point of intersection of that path and the plane constitutes a curved surface, a surface with a depth: a scenography. According to Leibniz's explanation, everything unfolds as if the order of iconography finally required the order of scenography—"but the same order demands that the effect be obtained" by which that order is removed. Everything unfolds as if iconography proved its own impossibility. That is why "Leibniz never saw the iconography. He probably proved that it was invisible. He never knew it, he proved that it was unknowable."[127]

But Leibniz's proofs were erroneous. Or, more precisely, his proofs were correct, but he started from an erroneous assumption: that there is depth and distance. Of course, if the existence of depth is unquestionable, then iconography must finally be proven to be unprovable. But Berkeley says that distance does not exist as far as the eye is concerned. This, of course, does not mean that in Berkeley's world there will be no experience of distance. It only means that other senses will constitute this experience. With respect to the eye, for its part, everything is absolutely close. Distance is not the experience of the eye. The eye is the sense of closeness. Closeness is an inevitable effect of the turbulent organization of the visible world. In this world, there is an infinite multitude of relative motions between relative spaces. It is therefore impossible to isolate two immobile places and a predictable motion between them. Relative motions intersect, redirect, and affect one another and act upon the body-eye with different "attractions and strikes," thus altering the position of the gaze of the body-eye, also. The eye, which is in constant relative motion, slides through the visible, affected by these attractions and strikes.

This is also how God's eye sees its own writing: as an infinite circulation of "pictures." In such a distribution of pictures, it is no longer possible to establish scenography, to isolate an immobile picture, and to determine a position from which it can be viewed: it is no longer possible to provide conditions for organizing the central perspective. God (as well as the finite eye) sees the world as a flat plane, as an endless surface made of relative spaces—places, or, as an infinite wall in which every "brick" is a *locus*. But, of course, this wall is not petrified. It is a wall in which bricks change their places. It is a living fluctuation—*fluctuatio animi*. God's gaze sees the world as an infinite living mural, and each picture is a flat spot that slides along the mural precisely in the same way as in fresco painting or in a Byzantine icon, where the surface (the wall or the wood) is both "the ground," the architectural foundation, and the "plastic form": both the background and the figure. What for central perspective has to be a "categorial scandal" (the indiscernibility of foreground and background) now appears, with the absence of the depth, as the only possible "appearance" of the visual field, a field on which "pictures" slide like on a smooth plane constituted by ideas themselves. The fluctuation of every "picture" presupposes and produces the fluctuating gaze.

God's gaze is dynamic. It is not a gaze that stares from its determined point of view, but one that is disseminated all over the infinite surface, a gaze that is "dispersed," together with pictures, in all directions, perceiving them simultaneously and everywhere, from all profiles, from all sides, thus, through this perceiving, maintaining their existence: it is the gaze itself that becomes this infinite fluctuating plane. God's gaze *is* the infinite mural of the world. It is a gaze that is absolutely close to every picture because distance does not exist in itself: "distance, of itself and immediately, cannot be seen." It "projects only one point in the fund of the eye, which point remains invariably the same, whether the distance be longer or shorter."[128] The gaze always sees two-dimensional spots, and because distance itself does not exist, they are always spots in the eye, or, conversely, the eye on the spot. God's gaze sees iconography.

Iconography is "the sum of horizons,"[129] which emerge from the background onto the "foreground." It annuls the difference between foreground and background, figure and its ambient background. The visible world is seen according to the logic of a background that is on the foreground and a foreground that "falls" into the background: on a flat, two-dimensional plane, there is a "flat" two-dimensional profile. This is what Beckett, in his "interpretation" of Berkeley, represents by the wall and the object of vision (O) on/in this wall. This object is not "stitched" to the wall as a foreground that makes of the wall a background. On the contrary, it is a mobile spot on the wall, a colored spot on the background, yes, but also the background on the colored spot. The visual field is abolished that was organized according to the logic that holds that by perceiving, the visible perception "frames" what it perceives and imposes this "frame" (or a form) as an "empirical" limit to the perceptual field. There are no frames anymore. Within the "logic" of iconography, "the figure loses its logical status as that object in a continuous field which perception happens to pick out and thereby to frame; and the frame is no longer conceived as something like the boundary of the natural or empirical limits of the perceptual field.... Whatever is *in* the field is there because it is already contained *by* the field.... It is thus the picture of pure *immediacy*."[130]

Iconography is the totality of all the profiles of the world, seen in their immediacy, or, iconography is the world seen by an innocent eye, which does not recognize distance or the projection of the vantage point. God's eye is innocent. It does not grasp its own text through a mediation that, as in Leibniz, presupposes a vertical axis along which the gaze distances itself. In other words, if God sees the world as iconography, it is because he does not have a reflexive form of vision that would double vision by doubling its axes, giving it an "apperceptive" form. God does not stare at himself. The gaze of the Berkeleian God does not see itself. God's vision is blind. Or, his gaze is not blind insofar as it perceives the world, and it is blind insofar as it does not see that it sees. God's eye—we will use Beckett's solution here—is therefore similar to a camera that is forever rolling and behind which there is no other eye looking through it—there is only the lens, which registers flat planes within an angle smaller than 45 degrees and objectifies them. "E is the camera." But no one stands behind this lens. God is the lens that objectifies. The work of this lens is the condition of possibility for objectivation, for the existence of the perceived—*esse est percipii*. But this perception is not perceived. Objectivation does not produce "subjectivation." Everything unfolds as if the object objectifies the objects of vision, as if there is no subject. There is no gaze to accompany God's gaze.

This is not to say that God does not have a mind, or that he does not have eyes. On the contrary, God is mind and God is eye, but this mind is not "behind" the eye, nor is the eye "an external" organ of the mind. Only in this way can we understand the two Berkeleian theses that at first sight appear contradictory: one that says that "the eyes of the Lord are in every place,"[131] that God constantly perceives and that therefore God has eyes, and the other that says that God is pure mind, that his ideas "are not convey'd to him by sense," and that therefore God has no eyes. These two theses are not contradictory. They assert the same thing: God's mind is his eye and God's eye is his mind. There is no eye behind the eye, no gaze that looks at the gaze: no apperception. God sees all and knows all, but his seeing is nonseeing and his knowledge is nonknowledge—God perceives us constantly, but does not know it. That is the way in which Berkeley tried to "remove" the blind spot of Leibniz's panopticon, according to whose logic there should exist absolute visibility and knowledge of this visibility.

The world is no longer organized, as in that logic, as a panopticon that overlooks the "surplus of gaze," the fact that there is always a blind gaze escaping apperception: apperception can see itself and its object of vision only if it does not see the gaze that sees it. The only way to subvert the blind gaze that stares at apperception "from behind" is by negating the distance between gaze and vision: perception and apperception fall into one, and the entirety of apperception becomes perception. There is no longer any distance between the eye and the visible. The visible is within the eye, and the eye is within the eye—a pure and absolute visibility of which one cannot but remain ignorant. The blindness of the panopticon is sacrificed for the innocence of an immediate vision that is blind. The blindness of mediation is exchanged for the blindness of innocence. God, who constantly perceives us, whose persistent gaze maintains us and who, in this way, produces our "most absolute and immediate dependence on Him,"[132] appears in his absolute indiscretion as an absolutely discreet God who does not notice us at all, even though he constantly perceives us. This God is, therefore, a degenerate Argus who neither oversees nor inspects, who is not a detective. He explicates himself through his visual language, which is an outcome of a "primary process" that does not grasp itself, that produces its effects "blindly," an unconscious process. In his rebellion against atheism, Bishop Berkeley thus revealed the true formula for atheism. "For the true formula of atheism is not *God is dead* . . . the true formula of atheism is *God is unconscious*."[133] We are in the "most absolute and most immediate dependence" upon an innocent, blind God.[134] But we do not know it.

The only thing we know is that we are continuously perceived, that we are always in the field of perception of God's eye, and that we cannot escape this perception because we exist only as long as we are perceived. And it is precisely this knowledge, which knows that we are exposed to the gaze of God, but does not know that his gaze is blind, that "fills our hearts with an awful circumspection of *holy fear*." Holy fear is the horror caused by the "feeling" that God's gaze is "with us wherever we go," haunting us, thus making our situation absolutely unbearable because it does not let us go. It does not allow us to hide. We are always at its disposal, transparent and visible. This gaze, which always successfully pursues us, which is curiously and precisely and always where we have hidden, provokes in us a continuous "agony of perceivedness," as Beckett would say.[135]

It announces the necessity of being with and within the Other, the impossibility of being absolutely alone. It is this inescapable burden of the constant presence of the Other, the situation of constant perceivedness, the impossibility of loneliness, that is intrinsically horrible.

This horror is what, in Beckett's interpretation of Berkeley, leads the subject of the Berkeleian world to attempt to hide himself from E, from the pursuing eye of God. This subject, who "had exhausted 'all joys of *percipere* and *percipi*'"[136] is trying to free himself from perceivedness by "hugging the wall on his left," the wall of iconography, with his "right hand shielding exposed side of face." However, his attempt to escape perceivedness is futile, since "the unprotected side alternates relative to the angle of view."[137] Every side is always unprotected, because God's gaze is everywhere. It does not recognize the angle of immunity. It is not the perspectival gaze, it is the iconographic gaze deprived of the distance from the object—the object of vision exists only in this eye, visible "from all sides," from all angles, from all profiles. In short, it is impossible to escape holy fear. But "holy fear" is an effect of a terrible misunderstanding, an outcome of the complete ignorance of the fact that God sees us, but knows not that he sees us, that we are constantly unseen, in spite of the fact that we are constantly perceived. We do not know that God's eye is blind. "Our" gaze is, therefore, a gaze that is constitutively blind to the nature of God's gaze. This is the entire paradox of the Berkeleian idea of constant perceivedness: the blind gaze of God looks at a gaze that is blind to the gaze of God. The horrible agony of constant perceivedness is eternal.

PERSON

The Room

Exhausted by being constantly perceived, O is trying to escape the gaze of E and to hide behind the door of his own room. He is trying to reach the invisibility of interiority. He is trying to leave the iconographic surface of the wall without depth and to "disappear suddenly through the open housedoor on his left." He mounts the stairs hastily, trying to become invisible, but:

> E transfers to where O last registered. He is no longer there, but hastening up the stairs. E transfers to stairs and picks up O as he reaches first landing. Bound forwards and up of E who overtakes O on second flight and is literally at his heels when he reaches second landing and opens with key door of room. They enter room together, E turning with O as he turns to lock the door behind him.

Exteriority has entered interiority. O is not preserved from perceivedness. By entering his own interiority, he has entered percipi. *Perception becomes dual, and "O is always seen from behind." O realizes: In his interiority, in his room, there is an external world—"Side by side on floor a large cat and small dog. . . . Cat bigger than dog. On a table against wall a parrot in a cage and a goldfish in a bowl. . . . mirror; window; couch with rug; dog and cat staring at him; parrot and goldfish, parrot staring at him; rocking-chair; dog and cat staring at him," plus photographs and a print of the face of "God the Father." O realizes that he perceives all those things and that exteriority exists insofar as he perceives it. In his eye, in his mind, there is an exteriority that makes of his mind an external mind. O tries to make this exteriority invisible. He "approaches window from side and draws curtain. . . . Holding rug before him he approaches mirror from side and covers it with rug. . . . He tears print from wall, tears it in fours, throws down the pieces and grinds them underfoot." However: the rug is now staring at him, the wall is staring at him. He is still in the field of percipi.*

Desires and Powers

On September 10, 1792, Samuel Johnson writes a letter to Bishop Berkeley in which he confesses to him. "I must confess," he says, that after reading Berkeley's books, he is "almost convinced" that matter is nothing other than "a mere non-entity." He also must confess that he is "almost convinced" about many other things that Berkeley "so nicely puts." He is "almost convinced" that the *esse* of everything that exists is that it is perceived. He is nearly convinced, as well, that all our ideas are inert. And yet, Johnson also "must confess" that there is one thing regarding which he remains completely unconvinced. This thing is not only unconvincing, it is also absolutely shocking, and not only for him: "It is after all that has been said on that head, still something shocking to many."[1] You claim that everything in the world, and, therefore, the human body too, is but an appearance, a ghost, "*nothing but a mere show*," and that behind this show, behind this theatrical play, there is something that is not represented, something unchangeable. All things in the world exist, therefore, as representations of unchangeable ideas, of the essences that are in God's mind. "Now I understand you, that there is a two-fold existence of things or ideas, one in the divine mind, and the other in created minds." I understand that you have twofolded the world, as if there are two worlds, one of ours and the other of his, the world of archetypes with which this world of ours is more or less in accordance. However, if you allow the twofolding of the world, then you are in a kind of disagreement with yourself: "You say the being of things consists in their being perceived. And that things are nothing but ideas, that our ideas have no unperceived archetypes, but yet you allow archetypes to our ideas when things are not perceived by our minds; they exist in, i.e., are perceived by, some other mind."[2] You claim, therefore, that things are both perceived and not perceived, that they do not have archetypes and that they have archetypes, that they are all in God and that they are not in God at all, because God has his own world, and, in general, you claim nothing but contradictions. You must answer me, therefore, regarding all this, and elucidate these contradictions.

Berkeley answers two months later and tells Johnson that he has had terrible headaches, which, of course, "very much indisposed him." Then he mentions the problem of natural philosophy, which Johnson had casually mentioned. He speaks, also, about the problem of guilt and punishment, but does not ever mention what was shocking to Johnson: the problem of archetypes. Instead, he has something to recommend to Johnson: "I will venture to recommend . . . (1) To consider well the answers I have already given in my books to several objections."[3] I would like, in other words, to recommend that you carefully read what I have written and think about what you read. Besides that, I do not have anything else to recommend. Of course, "My humble service to your friends," and as for you, "I am your faithful humble servant."

Johnson answers: I have read again and carefully all the answers you have given in your books to several objections, and I have found nothing about the problem of the twofolding of the world, which twofolds itself necessarily in the moment when you establish the existence of archetypes as the existence exterior to "our" world.

> And this exterior existence of things (if I understand you right) is what you call the archetypal state of things. From those and the like expressions, I gathered what I said about archetypes of our ideas, and thence inferred that there is, exterior to us, in the divine mind, a system of universal nature. . . . I humbly conceive you must be understood that the originals or archetypes of our sensible things exist in *archetypo* in the divine mind. The divine idea, therefore, of a tree I suppose . . . must be the original or archetype of ours, and ours a copy or image of this. . . . When, therefore, several people are said to see the same tree or star, etc. . . . it is (if I understand you) *unum et idem in archetypo* tho' *multiplex et diversum in ectypo*.[4]

If I understand you right, you claim that the divine world is an infinite world of essences and that the phenomenal world is only the deceptive form in which these essences appear, that it is multiple, movable, temporal, and changeable, a mere false copy of the truth.

The Passive Synthesis of Exhaustion 99

After this double insistence, Berkeley answers Johnson in two sentences: "I have no objection against calling the ideas in the mind of God archetypes of ours. But I object against those archetypes by philosophers supposed to be real things, and to have an absolute rational existence distinct from their being perceived by any mind whatsoever; it being the opinion of all materialists that an ideal existence in the Divine Mind is one thing and the real existence of material things another."[5]

If Berkeley has no objection against calling the ideas in the mind of God archetypes of ours, what, then, does he actually reply to Johnson? He replies: I do not mind if you call ideas in God's mind archetypes, but I do mind if you call them real things, if you hold that only what has absolute existence really exists, namely, has the existence that only God can perceive. What you call an archetype can exist in the divine mind, but not in the way of absolute existence. Therefore, it cannot exist as an archetype. Indeed, I do not mind if you claim that archetypes exist as archetypes, that they exist in the divine mind, but I do mind if you claim that they exist only for God, and therefore archetypally. To put it simply: archetypes exist on condition that you do not comprehend archetypes archetypally, for the existence of archetypes is not archetypal. The existence depends only on whether it is perceived by someone, and someone means anyone, not only the divine mind, but also any finite mind. Only materialists think that archetypes have ideal existence, meaning absolute rational existence. Only they think that archetypes exist archetypally. But this materialistic conception is untenable. From the archetypal existence of archetypes it should indeed follow that there exists one and the same divine world—the archetypal one—which copies itself into a multiple and different world of copies. If we were to claim that there existed two worlds, we could never, most importantly, comprehend a single reasonable reason for the existence of this other world of copies. If the whole universe exists archetypally in God's mind, why should God create the world of copies? And why should he maintain the copies in their falseness, if he is able to observe and maintain the existence of the truth, the archetypes? Even if, for some unattainable reasons, there existed two worlds, it would mean that we, in this world, perceive only copies, that we are deprived of truth, and that skepticism is the only true philosophy.

A hypothesis about the archetypal existence of archetypes is, therefore, a supposition that would refute the original motive of Berkeley's philosophy: the refutation of skepticism that lives on the idea of a twofolded world. That is why when it comes to archetypes, Berkeley claims exactly the opposite: the nonarchetypal existence of archetypes. Archetypes do not exist in the manner of ideal things. They do not exist in the way of "real things which have absolute rational existence." On the contrary, they exist in the manner of ideas, in the manner of "real things" that have, as all real things, relative existence. They have, therefore, not "rational," but sensible existence, which consists in being perceived by some mind—any mind at all. Archetypes exist as real, sensible things.

But, if archetypes are ideas or sensible things, why are they called archetypes? What is the difference between non-archetypal and archetypal sensible things? Or, because the real, sensible thing is nothing but a body, what is the difference between the archetypal and non-archetypal body? Archetypes or archetypal bodies are powers. The archetype is potency in the divine being. This is not to say that the archetype or potency has no existence or that the existence of potency is "ideal." On the contrary, potency exists as a thing that is a power—as a sensibly perceivable potential thing. The paradox inscribed in the archetypal (that it is the reality of a potential body) derives from a potency that is nothing other than desire: "If anything is meant by the *potentia* of A.B. it must be desire."[6]

The archetypal body or the power determined as desire means both that desire exists "sensibly," that it exists only if it is sensed, and that desire does not exist in the manner of any other body: desire cannot be touched or seen. The body of desire does not have a form. Desire is not a bundle of sensations—that is the difference between the body of desire and any other body. Desire is a set of powers, the constellation of potencies. Desire is, therefore, a paradoxical body: both body and nonbody, it is the reality of a potential body. The infinite multitude of these paradoxical bodies called desires makes up "the unknown substratum," which is God. God, therefore, is not a substratum that supports desires. God is a substratum made of desires—desire is the substratum. Desires, these archetypal bodies, are not, however, in any way unchangeable essences. There are not two worlds. The body of desire does not reside in a place different from the place of the sensibly perceivable body. Desire is the power, as well as the body, activity, as well as indifference: "If anything is meant by the word *potentia* of A.B. it must be desire. but I appeal to any man if his desire be indifferent, or (to speak more to the purpose) whether he himself be indifferent in respect of w^t he desires till after he has desir'd it. for as for desire it self or the faculty of desiring, that is indifferent."[7]

Desire, which composes the divine being, is indifferent. It is, therefore, passive. As potential bodies, desires are in themselves indifferent. In themselves, they do not desire until the cause-reason of desire forces them into the activity of desiring. However, this interpretation is unacceptable, for it claims that the passive desire informs a passive God, whereas the nature of God is pure activity.

This is, of course, only another way of saying that passive desire cannot inform the divine mind: if potencies in the divine being had to "wait" for the cause of their own actualization, that would mean that God would have to wait for a God to create him. In other words, before desires, there would have to exist objects of desires that would actualize "their" desire. To suppose the passivity of desire is to claim that the absolute is deprived of activity, which is the condition of possibility for the absolute. To claim the passivity of desire is to claim that the divine being is lacking itself: that being lacks being.

We are thus confronted with a paradox. If being-desire does not lack being, one must suppose that what is at stake here is a desire that does not lack any cause-reason of the desire, a desire without deficit, which has to be passive because it has no reason to desire. Everything unfolds as if absolute activity manifests itself as absolute passivity, or as if absolute activity does not have the power of self-actualization. This paradox can be resolved through a turn in the determination of desire: desire is no longer conceived of as based on deficit. But even though it does not lack anything, even though it is passive, it is at the same time active, so that God is not a being whose own being escapes from it. There is a twofoldness of desire that means, first, that in the case of finite minds, desire is indifferent as long as it is not "activated" by the cause-reason of desire, and, second, that the split between desire and its object does not exist in the case of the infinite mind.

However, the twofolded determination of desire as desire separated from its object and as desire without deficit does not confirm Johnson's interpretation about the existence of two worlds. On the contrary, it confirms only that Berkeley knows something about the fundamental ambiguity of desire: "Sometimes we objectify it [desire]—and we have to do so, if only to talk about it. On the contrary, sometimes we locate it as the primitive term in relation to any objectification."[8] So desire is what is being objectified by the cause-reason that pulls it out of indifference and moves it to action, by means of which desire is supposed to objectify what objectified it, but desire is, at the same time, the condition of possibility for every objectification. The paradox of desire, therefore, lies in its primitivism. Conceived "primitively," in its existence, which enables every objectification, desire is passivity, but it is this passivity that is the very core of activity. This desire cannot be caused by anything because it is in the place of the cause of all causes. It is what causes itself. Therefore, it is not a matter of an activity that reveals itself in its nature of passivity. It is a matter of a passivity that reveals its disposition to activity. Desire is passivity that activates itself.

If desire is, as Berkeley claims, indifferent to "its" object to the extent that this object does not appear as the cause of desire, then desire is the potentiality that will remain in the "state" of actuality of potentiality as long as it does not objectify itself into its own object, and it will become power as soon as it objectifies itself, not through the mediation of some actual object of desire, but through the passivity of its own potentiality. If desire is indifference that makes itself "different," then desire is what produces, and what produces "reality"—what produces itself into an actual object. That is why God is not the being who lacks being. God is not desire that lacks the object of desire, because desire that produces itself as reality does not lack any reality: "If desire produces, its product is real. If desire is productive it can be productive only in the real world and can produce only reality. . . . Desire does not lack anything; it does not lack its object. . . . Desire and its object are one and the same thing."[9] Desire is the potency of its self-actualization into an object.

Thus the nature of archetypes is explained. Archetypes are one of two possible means of the existence of bodies: "Bodies do exist whether we think of 'em or no, they being taken in twofold sense. Collections of thoughts & collections of powers to cause these thoughts. these later exist."[10] In other words, bodies exist either as desires that cause themselves or as embodied bodies, both as powers capable of causing bodies and as caused bodies. The archetypal world and the sensible world are two different states of one and the same world. "Ideas of Sense are Real things or Archetypes,"[11] which is not to claim that all sensible ideas are archetypes, but rather that all archetypes are sensible ideas. It is always a matter of the same multiple world of sensible differences. One cannot say, as Johnson says, that the world of God is "*unum et idem in archetypo,*" although "*multiplex et diversum in ectypo.*" Quite the contrary, the world is multiple and different both in *archetypo* and in *ectypo* because archetypal bodies are also sensible ideas or things that can be perceived. And even if power were a simple idea, even if, therefore, there existed in the divine substratum only one power-desire, this substratum would still be multiple. Every simple idea is a complication of the multiple: "Is power a simple idea, seeing it includes relation?"[12] This question does not ask whether power is composed of differences. It claims that power is made of *relata* that remain different. According to Berkeley's understanding of relation, which claims that an "unclear, ambiguous concept of relation . . . deceives us," which claims that there are no necessary relations, it turns out that what is related through relation is not in the relation. The question "is power a simple idea?" therefore asks whether power can be simple in spite of the fact that it is multiple. That is why this question calls into question the very nature of the simple, which cannot be one, seeing that it is multiple. At issue is the possibility of reformulating the meaning of the simple: the simple is always infinitely complicated. Hence, the archetypal world cannot be one and the same. The one desire is the multitude of desires. The one, therefore, cannot be determined at all—it is an abstract idea, a fiction.

Everything happens as if God were a giant sum of desires that fulfill themselves in an enjoyment of self-actualization. The being of God is the being of enjoyment that enjoys itself through the objectifying of its own being (desires) into an "objective" being. Therefore, when it is implied that, for Berkeley, "God, and God alone, is the active cause who works in the 'material' world,"[13] it is said that desire alone is the active cause that "works" in the world. The world arises out of desire and out of enjoyment in satisfying desire.

Of course, "this cause keeps all things in existence and causes them to be what they are ... to act as they act."[14] At first glance, "causing" and "preservation" are not the same. To "cause" the world means to interrupt the existing "order," to change it, to produce something new. To "preserve" means to keep the already existing. To "cause" means to cause the vital principle. To "preserve" means to make the vital principle last, to prolong life: "For what is meant by being endowed with the vital principle, except to live? And to live, what is it but to move oneself, to stop, and to *change one's state?*"[15] But if to live, to prolong and preserve life, means to change the existence of what is caused, to change its condition, then "to preserve" means only to preserve the change. To preserve does not mean to keep the same, but to preserve the difference as difference. There appears a split between desire that objectifies itself and thus "kills" itself in becoming an object (for it is going to be changed immediately) and desire that is the will to be kept in life, a split between desire that buries itself, thus revealing that passivity is the truth of activity (for ideas are passive and inert) and the will of the substratum as desire not to fall into the death of "fulfillment." This is the basic situation of Berkeley's world: desire objectifies itself and becomes an object. The creation negates itself in the very moment of creation. What was created falls into passivity. If God were simply to create the world in a pure act, everything would fall into the absolute passivity and quietness of inert ideas, of satisfied desire. In order to keep the created in life, preservation is needed. What is called "preservation" is the infinite will to live. This will has the disposition of motion and is motion.

Horse and Stone

And vice versa, this motion is willful. It is the motion that wants to desire. It is a life that wants to live. However, this life is not the life of the body. It is not the life of what is created. Everything that is created, *natura naturata*, lives as what was created. What was created exists precisely as what was, lives as what was alive and as what has fallen into ruin as a lifeless object. The ruin is the consequence of the objectification of desire into its own object, absolutely satisfied, without desires. God creates ruins. The ruin, therefore, does not emerge as the consequence of the distance between "to will" and "to be able." It is neither an effect subsequent to creation nor a creation that came to nothing, for ruin exists. On the contrary, it is the climax of successful creation, which is simultaneous with creation itself: "ruin does not come after the work. In the beginning, at the origin, there was ruin. At the origin comes ruin; ruin comes to the origin, it is what first comes and happens to the origin, in the beginning."[16]

That ideas are ruins means that they exist just the way Berkeley claims they exist—as pure passivity, neither as death nor as life, but on the border between death and life: as the existence of the dead. Bodies are not quite dead death, but also not quite living life. They are passive life—life in suspension. This "paradox" of life would not exist if bodies were created as material bodies, if potencies were actualized in matter that does not perceive, that is deprived of sensations, already dead in itself.[17] In that case, the objectifying of life in Matter would mean its objectifying in something "not alive." But created bodies are not material, since Matter does not exist. The absence of Matter is the necessary effect of the nature of God, who is the mind made of potencies and desires that cannot create their absolute negation, Matter. This "cannot," of course, does not mark a deficit in God's power, for Matter is not something that could exist, but that God cannot create. On the contrary, Matter does not exist as potency, either. It is not what could exist as nothing, it is what cannot exist, even as nothing. It is not true that bodies are immaterial because the finite mind cannot perceive Matter. On the contrary, the impossibility of perceiving Matter is the effect of its nonexistence.

For reasons we might call ontological, rather than epistemological, bodies must, therefore, necessarily be immaterial. And everything that exists exists as immaterial corporeality, as collections of sensations made of minimal sensible elements, of pre-individual singularities. The same must go for desire, if the body is the self-actualization of desire. Desire is also a set of these pre-individual singularities[18] in their potency, and the actualization of a desire is at the same time the actualization of different virtual minimal "elements" of desire. Bodies are therefore made up of these elements as sets of sensations. Every body is sensible: "it is impossible for me to conceive in my thoughts any sensible thing or object distinct from the sensation or perception of it."[19] Everything that is created is sensible, but what is sensible is passive: "A little attention will discover to us that the *very being* of an idea implies passiveness and inertness in it."[20] The very being of what is created is passivity.

Following this roundabout, we return to the same path: passivity is the absence of motion and change that Berkeley identifies with life. What is created does not live, but exists. Everything happens as if life exhausts itself in an effort of self-embodiment. Passivity means that all possibilities are exhausted, that the power that has actualized itself cannot be actualized anymore. Through the self-actualization of power, what is possible is exhausted, created life is exhausted, and every body is an exhausted body. But what is this exhaustion? What does it mean for desire to be exhausted? Does it "exhaust the possible" because it is itself "exhausted," or is it "exhausted" because it "has exhausted the possible?" It exhausts itself "in exhausting the possible, and vice-versa."[21] The potency that has actualized itself exists as an exhausted actual. That is how the world of indifferent, passive bodies, the exhausted world, emerges. But in this way, a fundamental split reemerges: a split between active and passive, God and world, cause and consequence—an abyss that puts the cause out of reach.

Berkeley's God is not the immanent cause of the created. If God were the immanent cause of the created, what was created would also be active, by its own being active, precisely in the way that, in Spinoza, finite modes are not only the expression of infinite substance, but the causes of other finite modes. If, therefore, Berkeley's God were the immanent cause of the created, then sensations would not be passions, but actions. But if God is not immanent to what is created, does this mean that there is an inversion of the economy of desire, such that desire, in its potency, is not in deficit for its own being, and that only a realized desire, as the being that is "fulfilled," is the being that is in deficit for its own being? Both yes and no. Yes, because the very being of what is created is passivity, nonconfirmation of activity of the being of God. No, because passivity as nonconfirmation of God's being is not outside of this being. It is not his nonbeing. This "Both yes and no" is therefore only another name for the ruin, for what is neither alive nor not alive. Because God remains exterior to his consequences, Berkeley cannot, like Spinoza, say *omnia animata*. On the contrary, he must say that everything is unchangeable, since everything is passive.

But Berkeley does not say this. He does not say this because he introduces a reformulation of the efficient cause: the cause is exterior to the consequence in such a way that the consequence still remains within the cause, encompassed by it. Everything that is created differs from God. It is therefore "out of" his being, but also "in" God's interiority, In him we live, passively. In him we exist and have our being, as ruins. Everything that is created is out of God in an exteriority that is the interiority of God's exteriority. But this relation between inside and outside cannot be easily understood as a kind of "split" within an interiority produced by self-reflection. It is not a matter of an interiority that by means of self-mirroring objectifies itself into its own exteriority, because that would mean that the difference between what is created and the creator is not arbitrary and irreducible. But it is irreducible: by its very being, what is created is passive. By its very being, what creates is active. What is created is therefore the exteriority of an interiority that remains irreducibly exterior to this interiority—the cause is "out of" its consequences, but the consequences are in the cause. There are no indications whatsoever of Spinoza's plan of immanency. God is the pure motion of life, pure activity. Bodies remain passive, pure indifference, to be "affected" by activity. And absolute activity will keep in existence what is created. This means: God will maintain what is created, but what is kept in life will nevertheless never become alive.

God preserves what is created through motion (of desires, of life), which is his own motion. Only God is motion, and only he moves. Bodies never move. They are always passive:

> All that which we know to which we have given the name body, contains nothing in itself which could be the principle of motion or its efficient cause . . . on the contrary, if we review singly those qualities of body, and whatever other qualities there may be, we shall see that they are all in fact passive and that there is nothing active in them which can in any way be understood as the source and principle of motion.[22]

Bodies do not move, "there is nothing active in them," they are passivity "forced" to motion through an activity that does not belong to them. God moves a body: a body moves, remaining passive throughout all of "its" activity. "That the Mind which moves and contains this universal, bodily mass, and is the true efficient cause of motion, is the same cause, properly and strictly speaking, of the communication thereof I would not deny."[23]

Strictly speaking, the body rests in peace throughout all of "its" motion. All motions in the body and between bodies happen "against" the nature of the body. Such motions move the body without ever becoming the "interior" principle of the body. The motion of life is in the body, but this "in" remains outside: life does not "belong" to the body that it moves. And when the body is active, it is not itself active, but "something" else, different from it, is active within it and forces it to move and to live, whereas it remains forever passive. The body is a ruin even when it is alive. "The inert body so acts as a body moved acts. . . . But a body, inert and at rest, does nothing: therefore *a body moved does nothing*, . . . In fact a body persists equally in either state, whether of motion, or of rest.

Its existence is not called its action."[24] This is the determination of what is dead: what is dead is not what does not exist, but what persists in the sameness of itself, in passiveness, in the absence of every change that is life.

God keeps all life for himself. Bodies are always moved from a living "exteriority," which changes them, thus making of the universe a lively dance of ruins. This is, perhaps, the real meaning of Berkeley's thesis that everything is in constant flux and change. The change is forceful: pure activity strikes bodies, as Berkeley says, combining and recombining collections of them through "strokes." Thus, universal variation emerges, turbulence appears: It is true that for the actualized world, all possibilities are exhausted insofar as every individual body is exhausted. But what is exhausted nevertheless changes once "struck" by the pressure of change, which combines again what is exhausted by moving the exhaustion to exhaust itself, on and on, to get tired through exhaustion. What is tired is what "no longer has any (subjective) possibility at its disposal,"[25] and what is exhausted is what can no longer realize a single objective possibility, because no single objective possibility is left at its disposal. However, tiredness and exhaustion in passive bodies are one and the same: they have neither "subjective" nor "objective" possibilities—they are exhausted tiredness. Exhausted as it is, exhaustion is moved by a motion that is not exhausted at all and that combines and dissolves the bodies, as if the body that sits were forced to stand up and start to move or lie down.

It is not the motionless lying that is the state of ruination and exhaustion ("exhaustion does not allow one to lie down"),[26] it is sitting that is the state of exhaustion of the body, as Beckett shows in his "Film," where a person, completely exhausted by being constantly perceived, can only sit in the chair, not lie down. Lying is not passivity. It is either a "renovation," a gathering of strength, the opening of new possibilities, or nonexistence. Only sitting is the state of life without possibilities, the existence of exhaustion: "seated, without the strength either to get up or to lie down . . . forever."[27] The impossibility of "forever"—to get up forever, to lie down forever—is situated precisely in this "forever" of sitting, in the passivity that is so exhausted that it is indifferent to any getting up, to any lying down, that is so exhausted that it has no strength to live, no strength for nonexistence. That is why one cannot say that Berkeleian absolute passivity is reinforced by "secondary causation, the bodily process of grasping something or standing up and walking,"[28] because this standing up, this walking, even if it is determined as secondary causation, nevertheless remains the standing up of what is still sitting—what walks, stands up, or lies down is a body that has not risen from its chair at all.

So, thanks to this activity, bodies move and "come across" other bodies. The force of their encounter (the force of attraction or repulsion) determines the amount of sensible points that will "pass" from one body to another. This passing of sensible points is the effect of the immateriality of bodies: "By immateriality is solv'd the cohesion of bodies,"[29] because immaterial bodies can merge into a compact mixture that can be easily resolved through pure activity without introducing the problem of penetrability and elasticity. The problems of penetrability and of elasticity and the entire question of resistance disappear as a result of the nonexistence of Matter. In the Berkeleian world, therefore, absolutely all combinations between sensible bodies are possible, and there is no way in such a world to determine the "conditions of sensibility" for sensible motions "in order to exclude abnormal experience . . . and nightmares from the world of physical science."[30] There are no abnormal experiences. And precisely because changes that sensible motions produce cannot be determined through sensible conditions, they can be determined as surprises. The stroke of activity is a surprise for the indifferent body that has the "nature" of attraction and repulsion, but "the words attraction and repulsion may, in compliance with custom, be used where, accurately speaking, motion alone is meant."[31] In other words, motion is what is attracted or repulsed in bodies, and it affects itself by moving bodies.

However, this motion that composes bodies is not a mechanical motion. Even though mechanicists understand that the body does not move, but is moved, even though "mechanical philosophers . . . inquire properly concerning the rules and modes of operation alone," they nevertheless inquire completely improperly "concerning the cause"[32] of motion. They use the word "force" for motion "as if it meant a known quality, and one distinct from motion, figure and every other sensible thing."[33] But, indeed, the quality of what is "within" the motion but not the motion remains unknown. That is why, after all, we must say that this quality is the mystic one, and that is why we must reject its existence: "But what an occult quality is, or how any quality can act or do anything, we can scarcely conceive—indeed we can not conceive."[34] Such an occult quality does not exist, and this is the key reason that leads us not to accept the mechanicist view of motion.

But there is another reason for this refutation of mechanicism. Those who suppose the existence of a uniform, but occult force that moves bodies, those who suppose that "mechanical laws of . . . motion direct us how to act and teach us what to expect,"[35] deceive us, because they claim that there exist determined, predictable, and constant regular motions and that "there is therefore a constancy in things, which is styled the Course of Nature."[36] But that is not so: abstract force produces only abstraction. The constancy in things is a fiction. It is true that everything in nature is produced by motions, but motions are not constant, uniform, and regular. We cannot infer from a motion any law that would direct us and teach us how to act. Motions are multiple, and every motion functions according to its own law, is its own law: "Particular laws of attraction and repulsion are various. Nor are we concerned at all about the forces [of attraction and repulsion], neither can we know or measure them otherwise than by their effects, that is to say, the motions; which motions only, and not the forces, are indeed in the bodies. Bodies are moved to or from each other, and this is performed according to different laws."[37]

Whatever we might know of motion always amounts to the motion that is the multiplicity of its own laws. Whatever we might know amounts always to "particular" surprises, attacks, assaults, shocks. Everything is always unexpected. Nature is not styled—God does not care at all about "the style" of the book he writes. His visual writing is a flood of words, and this flood is what we call the course of things. There are an infinite number of laws that cannot be known in their totality. That is why "the mind should be fixed on the particular and the concrete, that is, on the things themselves,"[38] on each thing particularly. A given thing does not say anything about another. There is no inductive generalization.

Berkeley's person, therefore, is not Robinson Crusoe. Such a person could not use actions like those used by Defoe's Robinson in order to "discover" and to know an unknown island. Such a person could not have submitted itself to "happy" inductive generalizations. "Robinson Crusoe on his desert island, though he took care to salvage all available stores from the wreck, knew that to keep alive for any long time he should investigate the food resources of the island, more especially which parts of what plants were nutritious, which poisonous, which neither."[39] Robinson had to discover uniform and regular motions, to foresee encounters with other bodies, to find out whether they would be happy or unhappy, nutritious or poisonous.

The hypothesis is as follows: the fact that Robinson stayed alive proves the impossibility of the nonexistence of inductive generalizations. If Robinson had held to Berkeley's instructions about the necessity of fixing on the particular and "concrete," if Robinson had examined every plant particularly and started his research always anew, "out of nothing," he would not have survived at all. His investigation, even of one "limited set" of bodies such as is found on a desert island, would have lasted "hopelessly long." The limited set would be infinitely large in contrast to Robinson's finite lasting. But, and much more important, "had plants, men (and other animals) been random, casual aggregates instead of belonging to species, which are causal aggregates,"[40] not only would this island of "his" remain unknown to him, but Robinson himself would be at the absolute disposal of these random encounters with casual aggregates, which would poison and kill him.

Defoe's Robinson survives on the desert island precisely because he does not hold to Berkeley's, or, alternatively, to Hume's assumption about the randomness of the particular. The world in which Defoe places Robinson, is based, therefore, on two basic presuppositions: first, the presupposition that the world is organized by an order that connects all qualities and things into a causal totality in which certainly there are differences between what is encompassed by this totality, but whose differences are themselves linked in a well-ordered, predictable, and readable way. This world is organized, therefore, through a taxonomy that gathers these differences and links them with connections of similarity and causality that enable Robinson to establish, by enumerating, lining up, and classifying, "rudiments of several sciences about causal interrelations of many things."[41]

Defoe's Robinson starts with an assumption regarding a general science of order, the assumption that there are "arrangements of identities and differences into ordered tables." He does not, of course, use *mathesis,* because he does not classify simple natures, but because he does not find out simple natures, he does not "start" with a taxonomy applied to nature. He starts with an assumption that he uses only what nature uses, and that is taxonomy. He starts with an assumption that the very book of nature is a taxonomy, and he resorts to taxonomy as if *learning* from the book of nature, revealing an order in "complex natures (representations in general, as they are given in experience)."[42] The fact that the world is organized as an order of continuous differences, because "*taxinomia* also implies a certain continuum of things (non-discontinuity, a plenitude of being),"[43] enables Robinson to produce inductive generalizations. To find out what can be cooked, what one might drink, before he has cooked it, before he has drunk it, enables him, to read the world, to skip pages in this book, to read ahead in order to find out what will happen, and, as a result, to make of the world a safe place, a shelter that opens itself hospitably.

The idea that the world itself is organized as a taxonomy reveals the second presupposition of Defoe's Robinson: in order to discover the world as it is arranged in its taxonomic organization, it is necessary that the subject of knowledge already possess a certain knowledge of discovery. In a mirroring effect, he develops taxonomy as a science of "links and classes" in order to find out the already existing taxonomic nature of the world. Therefore, the question of whether taxonomy as science produces a taxonomically ordered world or whether the world, appearing in its taxonomical nature, produces taxonomy has to be a phantasmatic question. The figure of Robinson demonstrates the simultaneous bringing forth of the taxonomical world and of the taxonomical knowledge that grasps this world: "As for the absorbers, the plants, Crusoe required some knowledge of botany, at the farmers', gardeners' and housewives' level, as to what a plant species is and a plant-life history. Every specimen of one species . . . has many characters which distinguish it from specimens of other species and these characters are linked in an orderly way; the order is physical and causal."[44]

Robinson, as we see, does not need "a lot" of knowledge. The knowledge of a housewife or a gardener is enough for him. It is a little bit more than enough, because this little is also organized in a way that itself is taxonomical (the order and connection of ideas) and that, due to the nature of its organization, has the power to read the entire taxonomic book of the world. Of course, this does not mean that in this world that can be foreseen through reading, the subject is not in any way threatened. This does not mean that nothing creeps up on Robinson behind his back. It only means that Robinson can sense what is sneaking up on him precisely because what is "behind the back" is a "margin" of what is already seen and appropriated, what is tamed in a causal connection. What comes from "behind the back" therefore is never completely "behind." This "behind" always implies an "in front" that is "conquered," known, and that addresses Robinson "with the best intentions": "the world fills itself with well-intentioned mutter."

So the basic situation of Defoe's Robinson is revealed: it is revealed that Robinson is placed in the structure of the Other. The fact that Robinson is placed in the structure of the Other, although he is on the desert island, that he is found in the middle of that structure even before the appearance of Friday, is not at all paradoxical. The Other is neither the object of knowledge nor the subject that Robinson might encounter. The Other is neither the plant nor Friday: "the Other is neither an object in the field of my perception nor a subject who perceives me: *the Other is initially a structure of the perceptual field*, without which the entire field could not function as it does."[45] The Other is none of the particular things or persons that we might encounter or perceive. The Other is the connectedness of others structured within a unique field, with the consequence that, when we perceive the one clearly and distinctly, we perceive the others unclearly and indistinctly. When I see an image clearly, I also see, indistinctly, what is on its margin, what accompanies it. The Other is "the order of causality." That the Other exists means that "around each object that I perceive or each idea that I think there is the organization of a marginal world, a mantle or background, where other objects and other ideas may come forth in accordance with laws of transition which regulate the passage from one to another."[46]

That there exist others thus does not at all mean that Robinson lives in a world inhabited by other people. On the contrary, even if Friday had never appeared, Robinson would, all the same, have lived in the world of the Other, in the world of the neighborhood between things, a world that is structured a priori as the world of causal connections that is the condition of possibility for every taxonomy and for every induction. That is why the world of Defoe's Robinson is not a solipsistic world. On the contrary, it is not only a world in which the Other structures me, but also the world that protects me simply by appearing as an organized world in which it is possible to foresee the possibility of everything terrible, catastrophic, dangerous, and poisonous. "Filling the world with possibilities . . . and transitions; inscribing the possibility of a frightening world when I am not yet afraid, or, on the contrary, the possibility of a reassuring world when I am really frightened by the world; encompassing in different respects the world which presents itself before me developed otherwise; constituting inside the world so many possible blisters which contain so many possible worlds—this is the Other."[47] The Other is the reassurance of a causal connection. The Other is the condition of possibility for categories, a promise of truth, a gift of the power of calculation. Thus we might call this a priori structure of the Other the structure of the day, the structure of visibility and transparency, because it promises the visibility of what is not visible. It promises the transparency of the unclear, of what is on the margin.

But the Berkeleian person is not Defoe's Robinson. On the contrary, Berkeley's person is Michel Tournier's Robinson, who lives in a world without the Other, in a world in which there are only occasions, in which, therefore, something sneaks up from behind one's back and threatens without any possibility of foreseeing this threat. The world that I cannot determine as the place of safety in spite of being frightened, that I cannot determine as the form of shelter when something haunts me, because anticipations do not exist, such a world can only be called a world without possibilities. Being without possibilities does not mean that the actuality of the world cannot change. It means that every thing is in the moment of its origin, an "absolute" thing, without the possibility of connecting with the other, because there is no structure of the Other that conditions and guarantees the entire (perceptual) field as the field of connectedness of differences. An exhausted world—"A harsh and black world without potentialities or virtual ties: the category of the possible has collapsed."[48]

The collapsing of the category of the possible is another way of saying that there is no a priori network that organizes, connects, and hence sees what we do not see, what is behind our back, and what we could, if this net existed, see by looking at what we see. A simple example: as long as the structure of the Other exists with firm causal connections within the perceptual field, the perception of horror in the eyes of someone in front of us does not just suggest, but signifies that something horrible is behind us. Within this structure, between the view and the object of vision, there exists, therefore, the Third that connects differences. (Of course, the Third does not have to be a person, but anything that is causally launched into the causal chain of the visual field.)

Without the existence of the Third, one cannot count on the connection of differences. The existence of the Third constitutes the structure of the Other. Without this a priori network that sees what is not visible to us on our behalf, the view of the other does not protect us. It does not, by watching, render what it sees potentially visible for us, too. When the structure of the Other disappears, when the totality and connectedness of all the infinitely multiple points of view disappears, then there is a world in which only one point of view is left. This is the fundamental situation of Berkeley's person or Tournier's Robinson: "In Speranza there is only one viewpoint, my own deprived of all context. And this shedding of context was not completed in a day. At first, and as it were instinctively, I projected possible observers—parameters—onto hilltops or behind rocks or into the branches of trees. The island was thus charted by a network of interpolations and extrapolations which lent it different aspects and rendered it meaningful."[49]

It is this projected network of interpolations and extrapolations that constitutes the Other. The absence of this structure does not mean that behind the rock or on the top of the hill there is no other whom Robinson might encounter. It only means that every one of these others is without the Other, without the network of interpolations, without the structure, without anticipations. That is why in the absence of this structure there remains only one point of view. I see only what I see, without margins: "Today the process is complete. My vision of the island is reduced to that of my own eyes, and what I do not see of it is to me a total un-known. Everywhere I am not total darkness reigns."[50]

This does not mean that "next to" what I now see there is nothing. Rather, there is something that I do not see, and this something is not the indistinctness of the distinctness of the object of my vision, connected with it, but a completely different distinctness, in no way connected with the object of my vision. That is why only what is seen exists in the moment of seeing. The world of the Berkeleian Robinson is organized therefore in a manner that we might determine as the structure of the night: only what is lit by the lamp can exist and be seen. Behind that, there is a perfect ignorance, the existence of some absolutely unattainable world. From this darkness, anything can break out—anything does break out—and "strike" us with a stroke of surprise: "One might say that each thing . . . slaps us in the face and strikes us from behind."[51] This striking is possible because the world has been scattered into a multitude of unconnected things, into a world in which no kind of taxonomy is possible, in which there is no possibility of making a generalization on the basis of taxonomy. The effect of this impossibility is Berkeley's thesis that the mind has to concentrate on the particular: one must go from one thing to another, from one event to another, without the possibility of their connection. There thus emerges a kind of mad induction—induction without inductive generalization.

It is only from here that the second reason that must have guided Berkeley in his critique of abstraction becomes clear. That critique is based not only on the "fact" of the pure presentation of things, on a perception that is nothing other than what is perceived, so that there are no connections, only occasions. His critique of abstractions is also based on the impossibility of inductive generalizations. Robinson does not use abstractions: "I can only talk literally." In other words, if speaking still exists, it has to hold to Berkeley's proposition—a word for an idea, to every thing its unique and irreducible meaning. Any word that does not cover only one thing becomes an abstraction, a "useless and destructive luxury" that is to be denounced as mere prejudice.

We should remember the notion of depth as the essence hidden by the appearance of the surface, depth as the depth of the truth of the thing "itself." We should remember this notion, "which I never troubled to scrutinize" when I lived in the structure of the Other, because the Other is structured precisely as the organization of depth, as a scenography that refers to its margins. That is why this notion is not questioned in the structure of the Other. It acts as the other for this structure, which is why "I never troubled to scrutinize when using it in such expressions as 'a deep thinker' or 'a deep love.' It is a strange prejudice which sets a higher value on depth than on breadth, and which accepts 'superficial' as meaning not 'of wide extent' but 'of a little depth,' whereas 'deep,' on the other hand, signifies 'of great depth' and not 'of small surface.'"[52]

To unmask depth as prejudice or blindness means nothing less than to reveal the existence of a world in which there are no margins, in which there is neither background nor foreground, a world in which there is no perspective. To disclose the world as a world without the Other means showing it in the truth of its iconography, as a world of pure surface, as sliding on this surface from one object of vision to another in a random succession of others that "slap us in the face and strike us from behind." The world of Berkeley's Robinson is a world of pure surface, which is, of course, horrible, because everything in it becomes harmful, "because the world has lost its passages." In contrast to the blind faith of prejudices, depth is not at all the exposure to the terrible truth gained through the hard work of mediation. Rather, depth provides the ease of protectedness. Depth is established through a mediation between two extremes (of close and distant, of visible and invisible), a mediation that connects them through the "middle way," which is called the *medius*, the happy medium that reconciles differences. The horror of the experience of extreme differences is appeased by the work of the medium, which is *mediocritas*, the mediocre way.

In contrast to the blind faith of mediocrity, the surface becomes the place of weight and of all variety of horrors. There is nothing easy about the surface. On the contrary, the surface is precisely the place of constant catastrophe, of sudden ruptures or blows capable of annihilating all established configurations. The surface is nothing but the randomness of the breaking out of disasters, because "catastrophes are necessary for the production . . . of elements." The surface is the immediate wildness of these unconnected differences, which are not reconciled in a common truth.

It becomes clear now that if Berkeley's philosophy is solipsism, this is not because in the world painted by his philosophy the other does not exist, because there are no exterior bodies. On the contrary, the Berkeleian person is constantly exposed to exteriority, in constant encounter with others, and it lives as exteriority, in an exteriority without interiority. If Berkeley's philosophy is solipsistic, it is because it describes a world that is not organized by the structure of the Other, by a structure that would connect the perceptual field within a totality. Finally, to say that Berkeley's philosophy is solipsistic implies that it describes the world as surface: "the solipsistic sky has no depth." It is a sky of surface, without any middle way, a world in which the word *mediocritas* does not exist, a world of pure and horrible innocence in which every "object . . . collides with me with the force of the missile," by means of a motion that is never mechanical, foreseeable, or uniform.

Once activated by motion, passive bodies then suffer it. When passive and inert bodies are moved and encounter each other, the effect of the encounter is always suffering. When they touch each other, bodies suffer, just as they suffer when motion touches them. Bodies are always exposed to the touch of bodies that are in turn "impelled" to this "touching" by the touch of motion itself. It is never a matter of one body being passive and the other active. It is always a matter of the fact that in this "affecting," both bodies are equally passive, equally affected. Both bodies suffer motion that is not theirs:

> For if the true nature of things, rather than abstract mathematics, be regarded, it will seem more correct to say that in attraction or percussion, the passion of bodies, rather than their action, is equal on both sides. For example, the stone

tied by a rope to a horse is dragged towards the horse just as much as the horse towards the stone; for the body in motion impinging on a quiescent body suffers the same change as the quiescent body.[53]

However, it is wrong to suppose that this principle anticipates Kant's "third analogy." On the contrary, this is the principle that subverts Kant's third analogy of experience. According to Kant's analysis, community or reciprocity does not mean only the simultaneous existence of different "substances," but above all their mutual action: "It is absolutely necessary that all substances in the world of phenomena, in so far as they are co-existent, stand in a relation of complete community of reciprocal action to each other."[54] None of this is possible in Berkeley's world—stone and horse are not in a relation of communion because they are not in a relation of mutual acting. The basis of determination of the one is not in the other, but in the motion that influences both the one and the other. Bodies are not in communion because no body contains in itself the "causality of certain determinations" of another body.

Berkeley's bodies are not in dynamic community, in spite of the fact that the entire universe is a turbulence. Between bodies there never emerges *communio*. But the absence of community between bodies conceived as *communio* in Berkeley's case means also the absence of the relation of community conceived as *commercium*.[55] No commerce takes place between two bodies. There is no commercial circulation, no selling or buying, which are all dictionary "translations" of the term *commercium*. Between two bodies, there is no economy, there are no debts. Even the compositions and decompositions of bodies cannot be conceived, either as exchange or as gift. Life does not give life to bodies, and the understanding of nature "as the nature which gives" is suspended. Nothing of the economical paradise remains, nothing of the gesture of "winning paradise economically."[56]

Just as one cannot say that Berkeley's description of the relation between bodies anticipates Kant's third analogy, neither can one say that what is at stake here is a description of the world offered by an "obsessive visualist." It is not pure visualism to say "The passion of bodies, rather than their action is equal on both sides." In other words, one cannot say that "Berkeley is merely looking at his horse and stone, as Hume merely looked at his billiard balls," and that "direct empirical interaction between bodies" is missing.[57] Granted: interaction is missing insofar as no "action" is the action of the body itself, but, through what moves them, bodies encounter "directly and empirically," collide, or even break against each other or into each other.

The comparison with pre-Newtonian astronomy, which is usually introduced in order to infer and confirm the thesis of those who reproach Berkeley for his "obsessive visualism," thus fails. In the world of pre-Newtonian astronomy, "there are no causes, no effects, nothing active, nothing passive,"[58] in contrast to Berkeley's world, in which there exist both cause and effect, in which "the culinary fire" of desire is pure activity, in which bodies are pure passivity. In pre-Newtonian astronomy, "'heavenly bodies' display their heavenliness by being visible but, in principle, intangible,"[59] but they are intangible exclusively for some presupposed Third who watches them and to whom, after all, they merely display themselves as heavenly. This is what the assumption about Berkeley's "obsessive visualism" overlooks—that Berkeley does not start from the position of the Third, that he speaks, as it were, from the place of the body itself, and that such bodies, one to the other, are never heavenly or intangible, that they touch and strike and disperse into a dust that blends with other bodies. Berkeley's thesis that bodies always suffer is not the judgment of the visualist who watches two bodies from a distance, the body of a stone and the body of a horse, or the bodies of two billiard balls that collide and return and hit the cushion. Moreover, it is not at all a matter of judgment. It is the recollection of a sense through which the body of the one who describes has passed, an uncertain description of a certainty of sensation on the part of the one who is, in "his" passivity, forced into motion. Stone and horse, mineral, billiard ball, and organism suffer equally. What is alive and what is not alive suffer equally, because as bodies, they are all equally not alive.

But if bodies are passive and indifferent, how is it possible that they suffer? How is it possible for there to be an indifference that would be passionate, but still indifferent? Clearly, this is no paradox. There is nothing paradoxical in the fact that ruin, understood as the passive life, is exposed to blows and endures them. The blow, of course, is always a touch. The stroke is an activity of touching, but the activity of touching is not the only possible way of touching. That is to say, the body that suffers the touch is the body that touches through being touched and not by acting at all. The touch, in other words, does not always have to be conceived as percussion, as an active force that exerts pressure on the passive body that endures it. The touch can, also, be nonactive, pure, passive transitivity, a passage for what acts: "the stone, without doubt, does not 'handle' things . . . but it does touch—or it *touches on*—with a passive transitivity."[60] Through this passive transitivity, the activity of touch is performed. Through it there passes the rude "entelechy" of the pure activity of desire that could bring about the catastrophe of the touched. The moment of catastrophe does not have to cause suffering—the catastrophe does not have to be catastrophic. The blow that finally causes the body to disassemble in fact pronounces the end of every suffering for that particular body: "Pain and suffering begin with existence and end when it ends."[61]

Pain, in other words, begins with the self-fulfillment of desire that has actualized itself and thus exhausted itself. The existence of suffering, therefore, is nothing other than the existence of exhausted possibility. And it is precisely from here, from this pain of actualized desire, that the whole truth of passivity and indifference appears. Indifference does not refer to the idea that there is nothing to be sensed. On the contrary, it means that there is nothing to be sensed but the pain of exhaustion. Exhaustion is exhaustion from pain. The only way for the exhausted body to escape constant suffering is in its deactualization.

Deactualization is what Berkeley's person-body attempts to actualize in Beckett's interpretation—to remove everything that might force it into action:

> He sets down case, approaches window from side and draws curtain. He turns towards dog and cat, still staring at him, then goes to couch and takes up rug. He turns towards dog and cat, still staring at him. Holding rug before him, he approaches mirror from side and covers it with rug. . . . He turns towards dog and cat still staring at him. He puts them out of room. He takes up case and is moving towards chair when rug falls from mirror. He drops briefcase, hastens to wall between couch and mirror, follows walls past window approaches mirror from side, picks up rug and, holding it before him, covers mirror with it again. He returns to briefcase, picks it up, goes to chair, sits down and is opening case when disturbed by print, pinned to wall before him, of the face of god the Father, the eyes staring at him severely. . . . He tears print from wall, tears it in fours, throws down the pieces and grinds them underfoot. He . . . gets up, takes off overcoat, goes to parrot, close up of parrot's eye, covers cage with coat, goes back to chair . . . and is . . . disturbed by fish's eye. He . . . gets up, goes to fish, close-up of fish's eye, extends coat to cover bowl as well as cage, goes back to chair.

He does so in order to suspend every activity, in order to soothe even the rocking chair, in order, finally, to release all exhaustion in "slumber": "Chair is soothed, arm is resting." The exhausting of exhaustion is suspended.

But this suspension is deceiving. Its possibility is not actualized, even in Berkeley's interpretation, until the final blow falls, until the dispersion of what is exhausted begins. It is only this blow that will bring the suspension of all exhaustion. The blow as the climax of pain announces the suspension of every pain. The disposition of this blow is, therefore, mad: it is pain taken to its culmination, the excess of suffering that appears as pure joy—pain releasing itself from itself. The body is struck by an unbearably painful blow, which is the blow of release that releases it from any further compulsion to move. The body disappears through a joyful disappearance, thus finally entering "the element which moves," the pure motion without compulsion, without suffering, the pure, incalculable motion of desire that is life.

It becomes clear why Berkeley claimed that his doctrine puts "to the ground" Spinoza's philosophy.[62] Spinoza's *Ethics* is based on the idea of increasing joy and reducing sadness. Absolute joy, of course, becomes possible only if the individual is the adequate cause of its act: "I call that an adequate cause whose effect can be clearly and distinctly perceived through itself."[63] The adequate cause is the one that is affected by itself, the one that produces all its consequences from itself, which is to say that the only adequate cause is God. Only he knows absolute joy. All other causes are inadequate or partial. Their effect "cannot be conceived by itself without other parts." Therefore, all other causes suffer, because "we are said to be passive, when something happens in us of which we are only a partial cause."[64] The enterprise of Spinoza's *Ethics* was invested in the effort to reduce the amount of suffering of these inadequate causes through the "immanent" resuscitation of finite modes, through a procedure, therefore, opposite to that of Berkeley's, one that stands Berkeley's philosophy on its head. In other words, Spinoza tried to reduce the amount of suffering by making God an immanent cause of every finite mode, which thus becomes "immanently" alive, capable of acting. And this immanent life relates to every finite mode. It applies no more "to human beings . . . than to other individuals, all of which are animated, although in different degrees."[65] In Spinoza's solution to the problem, all bodies have the power to increase the "degree" of their joy. But they will never become adequate causes. They will never reach absolute joy. Simply put, bodies are given only the power to reduce the amount of suffering more or less, not to be released from it: "From this it follows that a man is necessarily always subject to passions."[66] No matter how greatly the joy is increased, finite modes always suffer. Human joy is always reduced by sadness.

Absolute release from suffering, however, becomes possible if Spinoza's plan of immanence is inverted, as Berkeley saw: "Spinoza . . . will Have God to be *Omnium Rerum Causa immanens* & to countenance this produces that of St. Paul, 'in Him we live' etc. Now this of St. Paul may be explain'd by my Doctrine as well as Spinosa's, or Locke's, or Hobbs's, or Raphson's, etc."[67] That "in Him we live" can, therefore, be explained following Spinoza, namely, that God is the immanent cause of all created nature. But this insight can be also explained by conceiving God as efficient cause, thus changing the meaning of Saint Paul's statement: created nature is the exteriority of an interiority, so that life is not the immanent principle of the body, so that bodies are forced to suffer constantly because they are inactive causes that do not do anything because they are made as powerless, sad bodies. And no matter how much joy is caused in bodies through the acting of the exterior cause, they will always "work out" the suffering. That is why bodies should be put in a state of passivity, complete indifference, pure suffering, which, taken to the extreme, disassembles them and releases them from suffering.

This is Berkeley's logic: suffering breaks the body, and through this "destruction" establishes pure joy, unspoiled by anything, absolute acting, pure life. After the "blow," the body will be disseminated into the "severe ocean" of which Berkeley speaks as the mixture of live differences. What was the sad, exhausted body now becomes motion. It is in the ocean or in culinary fire, in life itself, life itself. The body has lost its divisions and partitions, its organs and its juices, in order to enter life. The wall of Berkeley's room, which he speaks of as the partition between his working table and a tower on the other side of the wall, has disappeared. There is no screen anymore—the "room has lost its partitions and it releases an atom." It releases minimal sensible points into the "luminous void," into the ocean or fire, releases this "impersonal yet singular atom"[68] into the absolute acting of pure joy. That is why it can perhaps be said that Berkeley's philosophy is a form of vitalism. Passive bodies are only momentary, provisional islands that will be returned to the joy of the circulation of culinary fire. It is in this way that impersonal, pure life is obtained. That is why Berkeley's doctrine is Spinoza's thinking brought down "to the ground,"[69] an embodied Spinozism as the embodiment of pure joy.

Black Maelstrom and Vast Vertigo

To say that sensations (and also feelings)[70] are the only matter that exists in the world is to say that exteriority is formed of the sensible. This thesis would not be radical if it claimed that exteriority is made of sensations only for the subject that mediates it. However, this thesis becomes radical if it claims that all sensations are external and that there is no such thing as "internal sensation." This is precisely what Berkeley claims: "All ideas come from without, they are all particular."[71] Needless to say, this irreducible exteriority of sensations is the condition of possibility for a presentation without representation. Sensation is thereby posited as a "thing" in its immediate presence, untouched by mediation. In this way the Cartesian (and Hegelian) conception of sensation, according to which sensation is the effect of mediation of an interiority by an exteriority, is subverted.

In the catalogue of sensations that Descartes introduced, dividing them into the "perceptions we refer to objects outside us," the "perceptions we refer to our body," and the "perceptions we refer to our soul,"[72] the sensation is always posited as an effect of the relation of the soul with itself. However, if sensation "in itself" exists only as an unmediated exteriority, then it cannot be a form of subjectivation, the form of the self-relation of the subject, insofar as the subject always emerges only by feeling itself, by an appropriation of the "content" of the feeling that thus becomes its own content: "But if put in the feeling, the fact is a mode of my individuality, however crude that individuality be in such a form: it is thus treated as my *very own*. My own is something inseparate from the actual concrete self."[73] Thanks to self-appropriation, the sensation or feeling transforms the content that exists in itself into a content that exists for itself. It is a determination of "my own being." The "I" that feels that it feels what it feels also feels, through this feeling, the difference between itself and what is not itself, but by the same token, the "I" ceases to feel this difference as difference, inasmuch as every sensation (every feeling) becomes its own, inseparable from its "actual concrete self."

However, if sensation is comprehended as an exteriority, as something that is always already external, as never-represented presentation, then this exteriority does not have the form of self-feeling. It is not appropriated exteriority, it is the exteriority that does not belong to any interiority. That is to say, sensations exist, although they belong to no one. They simply are what they are, and they are always passive, inert things. But if sensations do not belong to anybody, and if the body is determined as an exteriority not mediated by any interiority, does this mean that there is no interiority at all, nothing that could determine this exteriority as its own?

To say that all sensations come from the outside is not to say that the understanding or the mind does not exist, but it is to say that the mind capable of producing itself by the labor of self-mediation is not presupposed. Or, instead of the mind that would assume the form of self-feeling, we are introduced to the mind that is not different from sensations. "The very existence of ideas constitutes the soul.... Consciousness, perception, existence of Ideas seem to be all one ... the Understanding not distinct from particular perceptions or Ideas."[74] "Soul," "consciousness," "understanding," and "mind" are synonyms, different names for the set of sensations. The mind is a collection of sensations or perceptions: "mind is a congeries of Perceptions. Take away Perceptions & you take away the Mind put the Perceptions & you put the mind."[75] As the "congeries" of perceptions that all come from the outside, the mind is pure exteriority. It is outside. *It is the body.* A radical consequence follows: if the mind is a bundle of sensations, if it is an already established constellation of perceptions, then it is an already established synthesis. The mind is not the force of action. The mind itself is not active. It does not perform the "active synthesis of reflection."

However, if the mind is given as already synthesized synthesis, and insofar as the transcendental is the effect of self-transcending that synthesizes itself, then there is no possibility for the production of the transcendental mind. To put it quite simply: the mind in Berkeley's philosophy is not the transcendental mind, if the condition of the transcendental is the activity of synthesis. And even when the transcendental is given as the transcendental apperception that precedes perception, as is the case in Kant, this givenness is also understood as the activity of synthesis. The transcendental condition is not only "the original and necessary consciousness of the identity of the self," but also the consciousness of an equally necessary unity of the synthesis: "in other words, I am conscious myself of a necessary a priori synthesis of my representations, which is called the original synthetical unity of apperception, under which rank all the representations presented to me, but that only by means of a synthesis."[76]

The transcendental subject, therefore, is not only a necessary consciousness of self-identity, but also the *act* of synthesizing that manufactures the subordination of differences to this "necessary" transcendental unity. But from such a determination of the subject it becomes clear that the subject as self-identity is the effect of a presupposed representation ("all the representations presented to me"). Representation is the foundation of the activity of active synthesis. Active synthesis synthesizes only representations. It presupposes an exteriority that is nothing other than representation, which is the effect of the mediation of the subject and therefore its self-representation. If the subject could not think its own identity without having in front of its eyes the representation of the identity of its action, then it follows that representation is posited as the effect of priority given to identity. Or, to put it differently, if self-representation as the possibility of identity is the effect of that identity, then identity is presupposed in the process of establishing an identity. Transcendental subjectivity produces itself only on the condition that it is already given. "There is one aspect, however concealed it may be, of the Logos, by means of which the Intelligence always *comes before*, by which the whole is already present, the law already known before what it applies to: this is the dialectical trick by which we discover only what we have already given ourselves, by which we derive from things only what we have already put there."[77] Identity before identity—that is the law of the transcendental subjectivity.

However, to say that the mind is already given as an exteriority, as a bundle of sensations, means not to presuppose anything that could or should be produced. There resides the radicalism of Berkeley's gesture: if everything is given, then there are no secret and presupposed "identities." But if there are no secret presuppositions, then absolutely nothing is given, because everything is given, because everything could be given. What is given is only the pure immediacy of a presentation that, in the absence of the mediation of representation, becomes both the "subject" to "whom" everything is presented and the "object" that presents itself: "By Idea I mean any sensible or imaginable thing."[78] The mind is a given constellation of ideas, an already established synthesis. This means that this synthesis is not the passive synthesis of contemplation insofar as it is carried out by the mind itself. The mind "produces" the passive synthesis of contemplation only on condition that it, being different from its object, does not produce anything, but surrenders itself to its object and falls into one with it. In the "logic" of contemplation, everything unfolds as if the mind does not have the strength to argue with the object in order to appropriate it, as if the mind is always already too tired to keep its distance from the object.

However, Berkeley starts from a passive synthesis that is already carried out. The "end" of contemplation (the falling into one of the mind and its object) is his point of departure: the fusion of the mind with sensation has always already happened. Berkeley's "subjectivity" is neither the outcome of the active synthesis of reflection nor the consequence of the passive synthesis of contemplation, but a pure product, a suspension of every synthesis. Not even the relations between sensations are the effect of the mind. Even they are considered passive: "relations are not doing the connecting but rather they themselves are connected,"[79] as if subjectivation had already occurred, as if every work were always already performed, as if the mind were so exhausted that it did not even have the strength to be passive. If, therefore, the active synthesis of reflection were to be determined as an always fresh, relaxed, working synthesis, if the passive synthesis of contemplation could be determined as a tired synthesis (for, "fatigue is a real component of contemplation"),[80] then this already performed and given synthesis could be conceived of as the passive synthesis of exhaustion, as a synthesis so exhausted that it does not do anything, a synthesis in which the mind does not do anything: "In this sense ideas are connected in the mind—not by the mind."[81]

The mind as exhausted synthesis, in other words, the "I . . . [as] not that to which our various impressions are related," but the constellation of these changeable "impressions," is what was recognized by Hume: "men are nothing but a bundle or collection of different perceptions, which succeed each other with an inconceivable rapidity and are in a perpetual flux and movement."[82] According to the presupposition of the "philosophy of reflexivity," it is necessary, in order to deduce the subject from this determination of man, to suppose the existence of pure understanding or "pure intellect," the existence of something that does not exist and that cannot even be understood: "Pure Intellect I understand not."[83] However, unless one presupposes such a thing as a self-identical power capable of subsuming to itself the differences of sensations, it is not possible to establish a causal relationship between the sensations, which means that there is no community of sensations. And "without community, every perception . . . is separated from every other and isolated, and the chain of empirical representations, that is, of experience, must, with the appearance of a new object, begin entirely *de novo*."[84]

Without apperception and causality, there is no continuity, either, and without continuity, the mind is only a theater "where several perceptions successively make their appearance; pass, re-pass, glide away, and mingle in an infinite variety of postures and situations."[85] The mind becomes the flow of unconnected sensations that is the absence of the subject and therefore the absence of the object. The absence of the object is, of course, an effect of the absence of recognition. As Descartes had shown, it is this presupposed identity of the subject that connects and "recognizes" the object by uniting different sensations into one:

> But what is this wax which is perceived by the *mind alone*? It is of course the same wax which I see, which I touch, which I picture in my imagination, in short the same wax which I thought it to be from the start. And yet, and here is the point, the perception I have of it is a case not of vision or touch or imagination—nor has it ever been, despite previous appearances—but of purely mental scrutiny (*solius mentis inspectio*).[86]

This mind that "from the beginning," before the sensuous encounter with the object, thinks the object, is the mind that connects what has been experienced as different (different sensations of the sight, touch, hearing) into the identity of an object. It locates the experience of the one sense as "identical" to the experience of the other sense, thus recognizing the object as one. If this mind is absent, then the object is no longer possible. This is the radical consequence of Berkeley's refusal of apperception: if there is no subject as an originating power (of self-synthesis), then there is no object either, insofar as the object = X is the correlate of an "I think" and insofar as "I think" is the condition of the object = X. But it is not just that Berkeley tries to reveal the secret of "reflexive philosophy," namely, that identity has to be established in order to be able to be established. The point is, also, that "reflexive philosophy" hides self-identity, presupposing it not to exist at all, unable to ensure it through the active synthesis of reflection: "Wherein consists identity of Person? not in actual consciousness, for then I'm not the person I was this day twelvemonth, but while I think of w^t I then did."[87] Apperception, therefore, misses precisely what it is supposed to identify, the "I" in the moment of apperception: "actual consciousness of oneself" is thus posited as actual unconsciousness of oneself. Apperception does not prehend the subject of apperception: the subject is "the blind spot of apperception." "Rationalism" thus makes use of a deceit, for it hides the fact that identity cannot be posited from a presupposed identity. It hides the fact that there is no identity, at least not in the sense in which it conceives identity—as a "synthesized" continuity of an "I": if identity means continuity and the possibility of the synthesis of that continuity, then there is no identity. This is Berkeley's fundamental thesis.

The mind is a nonsynthesized succession of sensations, a theater of percepts, as Berkeley says, and as Hume repeats. But Hume reminds us that the fact that he is comparing the mind to the theater does not mean that we should *identify* the mind with the theater. This comparison should not lead us to the conclusion that the mind is organized by the logic of scene and performance, by the difference between the viewer and the visible: "The comparison of the theatre must not mislead us. They are the successive perceptions only, that constitute the mind; nor have we the most distant notion of the place, where these scenes are represented."[88]

We do not have even the most distant notion of that place because the place of the event is not different from the event. The event is not "performed" at a place that is within a space. The event is the place of the taking place of the event. This is the meaning of Berkeley's criticism of "absolute space," the meaning of the thesis that there exist only "relative places," only "locations" within the event. Location is constituted through the event itself. The constellation establishes the location, and from the event there develops a relative space. Before the event, there is no space. Space, consequently, is not some kind of a priori form of sensibility. On the contrary, sensibility is an "a priori" form of space. Or, if we extend the theater metaphor, it would mean that sensibility is not a performance on a stage, but a performance without a stage. A theater without a stage should serve as a metaphor for pure presentation, in which point of view and object fall into one: "Each point of view must itself be the object or the object must belong to the point of view."[89] Or, in Berkeley's version, things are in the mind, and the mind is in things. Or, following John Locke, for whom the metaphor of a big house with many rooms and a veranda was closer than the metaphor of a theater, the veranda is all that exists of the house. There is no interiority to the house, only a noisy promenade on the veranda. Or, again, as

Locke says (enthralled by another metaphor), there are only characters drawn in the dust, only "grains" of dust connected "for a moment" by the activity of the wind. Locke identifies this "exhausted" synthesis "connected for a moment" with the "perishing of the soul," with the blind motion of animal spirits. "Characters drawn on dust, that the first breath of wind effaces; or impressions made on a heap of atoms, or animal spirits, are altogether as useful, and render the subject as noble, as the thoughts of a soul that perish in thinking; that, once out of sight, are gone forever, and leave no memory of themselves behind them."[90]

The "animal spirit" is given as pure sensibility, without continuity, memory, or past, without anything that is given besides the given, without the power of synthesis of the given. The animal spirit has nothing in terms of depth. It is not a "deep" spirit or a "profound" mind. It is a superfluous mind. It is nothing other than this always changeable surface on which the sensible is written, the pure life—the animal. But in saying all this about the wind and the dust, Locke is actually saying that the fact that this animal soul is superfluous, the fact that it does not have the depth of pure understanding, the fact that it is shallow or stupid, makes of it an ignoble being. The mind without self-identity, the mind of pure succession of differences, is a stupid animal mind, a bestial mind.

As is well known, this is not what Kant will discover somewhat later. He will claim, together with Locke, that sensibility "reduces man to a merely animal being."[91] He will identify pure sensibility with animality, insofar as pure sensibility condemns the mind to its givenness, thus depriving it of motives. But Kant will also claim that this given mind, precisely because it is given, cannot be evil or bestial, insofar as through the labor of evil, "the very opposition to the law gets the form of the motive." Far from being pure sensibility, evil resides in reason that is freed of the moral law and that, thanks to this "freedom," becomes diabolical. What is bestial, therefore, is not diabolical: according to Kant, the animal is not evil. However, Kant does not say that the animal mind is stupid, and this is not accidental, for at least in this case, Kant holds on to the logic of common sense, which fails to connect stupidity with pure sensibility and with the animal: "Stupidity (*bêtise*) is not, in principle, the character of a beast, *une bête*. In French, no one says of a *bête* (the animal) that it is *bête* (stupid). There are stupid beasts (*des bêtes bêtes*). . . . But, the stupidity of these beasts is a human stupidity."[92]

The stupid animal exists only as an effect of the projection of a specifically human feature—stupidity. Stupidity, therefore, is not given in the figure of the given, in the figure of the animal mind or exhausted synthesis. Stupidity is not given, it is, in fact, possible only as something that is immanent to the very process of individuation: "it is possible by virtue of the link between thought and individuation."⁹³ This is not to say that it is immanent to thought itself, but that it is the effect of the failed link between thought and individuation. In the very depth of the connection between thought and the labor of self-production, individuation begins to work "vacuously," not managing to form itself, not managing to give itself visible form. Stupidity is, therefore, a matter of a "depth" that does not succeed in "forming" itself. (Which is not the case with the animal spirit. Animal spirit, in its pure superfluousnesses, is always formed, although always differently.) What we see when we see a stupid person, then, is that there is nothing to see: maybe something is going on deep within this individual, maybe there is some link between his thought and his self-production, but it is not visible. Everything unfolds as if form separates itself from formlessness through thinking, but formlessness does not separate itself from form. Stupidity is "the indeterminate, but the indeterminate in so far as it continues to embrace determination, as the ground does the shoe."⁹⁴

"Reflexive philosophy," which is the term Berkeley likes to use, hides the fact that the reflexive subject has a privileged access to stupidity. "Reflexive philosophy" does not elaborate the immanent possibility of failed individuation, it always already presupposes the relation between thought and self-production and a subjectivation that is successfully carried out in the form of the subject in its unlimited will to truth. It comprehends the thought of the subject as the motion whose only conceivable failure is error. Clearly, error is a means of self-confirmation for reason and its power insofar as reason falls into error only through a gesture of erroneous recognition, through a minor confusion that is the effect of a moment of weakness. But error does not annul the reasonableness of the understanding. On the contrary, it confirms it: "Error, therefore pays homage to the 'truth.'"⁹⁵ Error denounces understanding that has failed in its effort to recognize the truth, but that sooner or later will identify its confusion and thus reveal the truth. It is understandable that within this "picture" of thought, there is no place for stupidity, which is, of course, stupid. For only in the space of the transcendental can the landscape of stupidity arise as an immanent and not accidental "scapeland" of subjectivation.

That "thought" is sensation, that one cannot establish the difference between thought and its object, does not mean that thought falls into a tranquil, idiotic identity. This sensible thought becomes what affects it, but because what affects it is always different, this thought is always a different "sensibility." It changes constantly by means of a change that is not the effect of its powerlessness to produce the "form" of its individuality (it is not a matter of a stupid mind), but the outcome of the pure difference of motion of "divine desires" that cannot be mediated by the reflexive gesture of synthesis. This is the entire point of Berkeley's thesis, according to which there is no identity if identity means continuity. There is identity only on the condition that identity means becoming something or somebody else. In other words, there is identity not if difference is assigned to identity, but if identity is assigned to difference: "My entire life, my memories, my imagination and its contents—all escape me, all evaporate. Unceasingly I feel that I was an other, that I *felt* other"[96] —always the continuity of a discontinuity, a blow and "intrusion" of a "new" sensation into a constellation. The new sensation changes a constellation into another constellation, changes the very "soul" of that constellation. Alteration of a set of sensations produces another mind. And because there is no distance between the sensible and sensibility, Berkeley cannot say, in anticipation of Husserl, that every consciousness is the consciousness of something. He can only say, in an anticipation of Bergson, that every consciousness is something.[97]

However, if consciousness is only a set of sensations, why does Berkeley still use the terms "consciousness" or "understanding" or "mind," since he already has at his disposal the term "sensation"? It is because this consciousness, which is something (the set of sensations), is something more or something else besides being this something. It is that something *plus* nothing, or more precisely, the not one, no one of the something insofar as this no one is not nothing, but something that exists as no one, as nobody, as a void: "And me, what is really me, I am the eye of all that, a center which has no existence, except that postulated by the geometry of the abyss; I am that nothing around which that movement rotates, without any justification except to rotate."[98]

"Consciousness" is the name that names both the set of sensations and this void, which is the only *res cogitans* that exists, which is the "*res reduced to nothing.*"[99] "I" is a no-thing of the collection of sensations in which everything that exists as something is exhausted, in which something that is not sensation, that is to say, *res* as *res cogitans*, can exist only as no-thing: "'to be' therefore, does not refer to a surplus in the 'form' of a substantially determined 'I,' for this verb 'to be' is, also, a white domino, a joker. Hand is no more a hand when it takes the hammer, it is the hammer itself, it is no longer the hammer, it flies, transparent, between the hammer and the anvil, it disappears and dissolves,"[100] like the mind that, in its transparent emptiness, flies in all possible directions at the same time, by means of a flight that nullifies any proper or appropriated identity, any proper name. Therefore: this no one is every possible one. It is namelessness, it is Ulysses, insofar as Ulysses is everybody, anybody, no one, capable of being everyone, no body capable of being anybody.

And Ulysses is anybody because he is blankness, pure openness—absolute and unconditional consent: "While I exist or have any idea, I am eternally, constantly willing, my acquiescing in the present State is willing."[101] "I" is nothing other than the force of "willing" of any name, the void that embraces any identity. That is why the "foundation of life" of this nameless Ulysses is a "no one," or "nobody" that, precisely because it is an impersonal consent, should not be named as "nobody" but as "yesbody." Yesbody is a nameless will that wills everything. Yesbody "accepts" being nobody in order to accept everybody. Yesbody enters into a floating life that knows damage of every kind, in which everything is possible because it is everything. "No one" is a thought without pure understanding. It is pure vertigo: "I think, I think endlessly; but my thoughts contain no reason. . . . My soul is a black maelstrom, a vast vertigo spinning around the void, a movement in an infinite ocean, around a hole in nothingness."[102]

Everything unfolds as in the logic of inverted perspective, where the position of the artist's gaze is not external, but internal to the painting. According to this logic, the artist enters the painting, "embodies" himself in it. The painting, therefore, is not the "internal" window open toward the world, but an exteriority from which all interiority has vanished: "the artist is not isolated from the world he represents, but places himself in the position of an observer involved in it, and his active transference within this world becomes possible."[103] The artist does not exist as the internal focal point of an exteriority because there is no bar that would delineate exteriority from interiority. The distance between center and periphery has disappeared: the center is the periphery, and that is why the artist can be "transported" into any periphery and become any exteriority. As if the subject had fallen out of its interiority into an exteriority. As if the subject had destroyed itself and become pure exteriority: "To create myself, I destroyed myself; I have so exteriorized inside myself that in my interior I only exist on the exterior."[104] Only an exteriority has remained, and it now "travels" from one object to another in a voyage that is the movement of transformation of one assemblage into another. Pure outsideness. Nobody can become everybody precisely because nobody is no/t/ one, because it is the multiplicity of transportations from one object to another, because it is the multiplicity of objects. We will call this subjectivity that disappears into the multiplicity of objects the "iconographic subject": "Objectivity is iconographic. The subject has disappeared."[105] The iconographic subject is the subject become "objects." It is an "objectivity" that knows itself as always-different objects.

But what does it mean that it "knows itself," that it knows its "own" existence? This knowledge of existence is not the knowledge of its uniqueness or of its continuity. Existence, comprehended as continuous "essence" mediated by accidents that can all be synthesized into a unity, does not exist. Or, it exists only as an abstraction: "This I am sure I have no such idea of Existence or annext to the Word Existence. & if others have that's nothing to me. they can Never make me sensible of it."[106] Being sensible is the only knowledge, but I cannot sense existence, I cannot feel being. I always feel only differences, the existence of what is at the moment, of what I am at the moment, and I am what I perceive at the moment: "Existence is percipi or percipere."[107] To be, therefore, means the sameness of perceiving and what is perceived—ideas exist only within the mind, but, as we have seen, the mind exists only within ideas. That is why "I think, therefore, I am" completely misses the point of the nature of subjectivity: "Cogito ergo sum, Tautology,"[108] because between "to think" and "to exist" there is no "therefore," no distance between the subject and the object: "I think" means I am what I think, whatever I think, and I always think something else: the mind is only a succession of perceptions. The I is no one. The only possible subject is the iconographic subject.

Let us repeat: even though the mind is a succession of sensations, it nevertheless knows its existences. Of course, this knowledge is not reflexion, but intuition. Through intuition, the mind instantly and immediately feels itself, not as something different from what it feels, but as precisely what it feels. Berkeley's determination of intuition is, therefore, very close to the way Spinoza defined reflexive ideas. Spinoza, too, could not have determined the reflexive idea as the idea of the idea, for that would have disturbed the parallelism of the successive finite modes of the two attributes. Spinoza found the solution for the problem of reflexive ideas in positing the reflexive idea as the form of the true idea that collapses into the idea that is the idea of the body and is united with it: "This idea of the mind is united to the mind in the same way as the mind itself is united to the body . . . For as soon as someone knows something, by that very fact he knows that he knows it, and at the same time he knows that he knows that he knows, and so on to infinity."[109] The whole infinity of a succession of reflexive self-apprehension falls into this "at the same time." It is compressed into a "moment" that negates the distance between knowledge and the knowledge of knowledge.

It was in order to negate this distance that Berkeley introduced intuition as the way in which the soul feels itself. The mind that perceives is the sensation that senses that it is what it senses. The problem of temporal distance, which was the key reason for Berkeley's criticism of the structure of self-reflexivity, should now be resolved. At any given moment, the mind senses what it is. The mind that feels pain becomes pain that feels itself, pure pain. If, therefore, it is at all possible to talk about the *cogito* of iconographic subjectivity, then that *cogito* can be determined only as the hypnotic *cogito*, as the *cogito* of a hypnotic trance, on the condition, of course, that "the hypnotic trance dissolves the subject, plunging it into a pre-representational state. . . . The hypnotized person has no 'self,' no 'ego'; not because that subject is divided or absent in relation to itself, but because it is so well enveloped in the 'here and now,' so very *present*, that it simply cannot be present *to* itself."[110] The iconographic subject is present precisely because the self-representation that produces "presence *to*" is absent. The iconographic subject is presentation without representation. Which is not to say that the mind-object is unconscious, but rather that it is a total consciousness: "In fact, this state, so often described as a loss of consciousness, could equally well be described as total consciousness, as a *con-scientia* so closely bound to itself, so present, that it literally short-circuits any self-distancing in which an 'unconscious' could lodge itself."[111]

It is true that Berkeley also establishes the difference between idea and notion: "We may not, I think, strictly be said to have an idea of an active being, or action, although we may be said to have a notion of them."[112] But the difference between idea and notion is nothing other than the difference between sensation and sensation that senses itself. If we understand the meaning of these words, and Berkeley persistently and explicitly insists on such an understanding ("inasmuch as I know or understand what is meant by these words"),[113] then we will also understand that the notion is not the notion of sensation. It is not a matter of the existence of an additional conceptual power that would accompany ideas and intuition. On the contrary, it is a matter of "conceptualizing" the notion as another notion for intuition—apprehension is only the immediacy of self-feeling. Berkeley tried to

> make his position rather clearer by coupling his denial that we have an idea of spirit with the assertion that we have a notion of it. As it turned out this was not a wise move. Rather than taking him at his word and seeing his point here as being that "we understand the meaning of the word," and that we know the meaning of the word because we have immediate and intuitive awareness of spirit through a reflex act of inner feeling, many commentators have supposed that for Berkeley a notion is itself a sort of thing distinct from whatever it is a notion of.[114]

But the notion is no such thing. It is intuition, which is immediate feeling: it is immediate consciousness, which is its object. In this way, two kinds of objects become one: objects that we perceive are nothing other than the "ideas actually imprinted on the senses,"[115] ideas that feel themselves in a kind of an "immediate" total consciousness. Only in this way is the main aim of Berkeley's philosophy realized: the restoration of everything to the state of an absolutely "objective" perception, which is the perceived, the restoration of everything to a state of "virginal" consciousness that is not separate from itself by any mediation, although it is penetrated by itself again and again. Everything is restored to a state without disguises or masquerades, a state of complete nakedness, complete visibility. The world without masquerade is a world in which identity is the proliferation of sensations and the intensity of those sensations. Instead of the multiplicity of representations, there is the multiplicity of presentations. The world becomes a world of "naked, undisguised ideas."[116]

Here emerges the central problem that Berkeley is trying to solve without at the same time subverting the possibility of immediacy and the nakedness of the things. The problem could be formulated—Berkeley formulates it—in the following way: is consciousness, or soul, or understanding, the same as the person? Berkeley's answer is "No, they are not the same" insofar as the person has in itself something that is given, but can nevertheless negate what is given. The "person" is at the same time both the exhausted object and the motion that Berkeley calls "Will." However, the will is not a substitute for the thinking thing. It is not the power of synthesis. On the contrary, it is substance without substantiality, without unchangeable essence. It is determined as the simultaneous movement of contradictory particular volitions that are not mediated into a unity: "The Will not distinct from Particular volitions."[117] And because it is not different from particular volitions, the will is precisely the difference between volitions: the willing thing is a multiple thing, the multiplicity of unconnected volitions. It is not one with itself, and therefore it is not a thing at all: "If you ask w^t thing it is that wills. I answer if you mean Idea by the Word thing or any thing like any Idea, then I say 'tis no thing at all that wills. . . . Again if by is you mean is perceived or dos' perceive. I say nothing wch is perceived or does perceive Wills."[118]

The "thing" that wills is not a thing because it does not perceive and cannot be perceived. There is no idea of the will ("certainly the Will is no Idea").[119] Will is not the object of knowledge: the will is what escapes us and what escapes itself, for in order to prehend itself, it would have to appear to itself as a thing, as idea. To put it quite simply, the circle is closed. The will is the unknown, the cause: "certainly the Will is no Idea, or we have no idea annext to the word Will. . . . The Will is purus actus."[120]

The will—which is the "total" of infinitely many different volitions—is, therefore, the "total" of infinitely many different pure acts. This is the whole nature of the person. The person, or the spirit ("the spirit" is the term that substitutes the term "person" in *The Principles of Human Knowledge*), is at the same time both the bundle of sensations, the understanding, a constellation of passive ideas, the object, *and* the imperceptible activity that acts within that object.

The Passive Synthesis of Exhaustion

The person is the mobile object. The identity of this subject, of the person, or of the spirit is the will that has no identity, that is the open set of different particular volitions traversing passive sensations: "Doctrine of Identity best explain'd by Takeing the Will for Volitions, the Understanding for Ideas. The difficulty of Consciousness of w^t are never acted etc surely solv'd thereby."[121]

The difficulty imposed by the passive mind or by exhausted synthesis is now resolved by the "injection" of divine particular volitions into particular sensations. The identity of the person means that the person is nothing other than the multiplicity of exhausted ideas in which there are volitions that move passive ideas, directing them in unforeseeable directions. Those particular volitions cannot perceive themselves. They do not know what they want, do not know anything about the directions in which they are directing ideas. "The Will is *purus actus*, or rather pure Spirit not imaginable, not sensible, not intelligible, in no wise the object of ye understanding, no wise perceivable,"[122] which means that the will cannot perceive itself, either, that it is a motion that blindly "acts, causes, wills, operates."

This blind activity is not accidentally named as "will." It is thus named because it is the blind motion of acquiescence in sensations: "Locke in his 4th book & Descartes in Med. 6. use the same argument for the Existence of objects viz. that sometimes we see feel etc *against our will*. . . . While I exist or have any Idea, I am eternally, constantly willing, *my acquiescing in the present State is willing*."[123] It is, therefore, not true that sometimes we see or feel, that sometimes we *are* against our will. Being "against our will" is a pure impossibility insofar as will is an unconditional consent, a constant "Yes." The person who sees is willing what he sees. He wants to be what he feels, whatever he feels. The person is always what he wants (to be), for the will is precisely blind consent, absolute acceptance. It is clear: this will is "given," but by means of what is "given," the "being given" of what is given is overcome. Because it is the simultaneity of the motion of "Yes," the will is an indefinite set of consents that are consenting to different sensations, and every "Yes" is a negation of an existing constellation and an affirmation of a different constellation, of a different "identity."

This is the transcendental principle of Berkeley's empiricism: the person is the impersonal motion of the transformation of an identity into another identity, blind consent to everything. The motion of the will is, therefore, the motion of individuation. Individuation is the effect of the force of the motion within the sensible. It is the effect of the intensity of consent. "It is intensity which *dramatizes*" an identity; "it is intensity which . . . determines an 'indistinct' differential relation in the Idea to incarnate itself in a distinct quality and a distinguished extensity."[124] The force of the blind motion of the consent is the very "labor" of dramatization that enables the differentiation of indistinct relations and thus establishes a new constellation, a new identity. The intensity of the consent is the force that forces the exhausted synthesis to "give" its consent to the new sensations. The fundamental principle of Berkeley's transcendental empiricism, therefore, reads as follows: The person is the iconographic subject moved by blind will into an endless repetition of the "Yes" that establishes new identities.

The Time of Pain and the Time of Waiting

If all objects of human knowledge, in accordance with Berkeley's classification, can be divided into "objects . . . actually imprinted on the senses; or else such as are perceived by attending to the passions and operations of the mind; or lastly, ideas formed by help of memory and imagination,"[125] and if the first two groups of objects can be apprehended as "one and the same" group, which can be differentiated within itself according to the presence or absence of the will within the objects, how, then, is the existence of the third group of objects possible? Or, to put it differently: how is the existence of memory possible, if memory presupposes the continuity of the person, which is, like any continuity, temporal? This question is urgent, since time is not the "formal condition *a priori* of all phenomena." It is not what within which "the whole reality of phenomena is possible," as Kant would have it. On the contrary, time is the effect of the "reality" of ideas. It is the a posteriori effect of the movement of sensations: "Time train of ideas succeeding each other."[126] Time is the succession of sensations, a succession that is discontinuous. There is no "simple idea" of time, or there is no time that could be abstracted from the motion of sensations: "Whenever I attempt to frame a simple idea of Time, abstracted from the succession of ideas in my mind, which flows uniformly and is participated in by all things, I am lost and embrangled in inextricable difficulties. I have no notion of it at all."[127]

The existence of time as the form of the sensible is thus unthinkable. There is no absolute time in which all beings could participate, or there are no two persons that "participate" at and in the same time. Where the structure of the Other does not exist, where there is no "network" that connects and mediates individualities, time cannot exist as a "common" form in which all beings equally participate. All that can exist are the different times of each person. Every person has its own time: the "time of every man is private."[128] Private time is not only the negation of Newton's absolute time, but also of "relative," common, or "public" time in which at least two persons can participate at the same time. All that is left is the existence of different times that are the speed of distribution of sensations. "Time a sensation, therefore onely in ye mind";[129] or, more correctly, time is the mind insofar as the mind is the bundle of sensations.

The speed of the movement of sensations slows down or speeds up time: "We are reminded of the lunatic logic of Joseph Heller's Dunbar in *Catch 22*, for Dunbar believed he could 'slow down time.'"[130] But the catch lies in the "fact" that Dunbar's logic is not insane at all, since it is not a matter of demonstrating that every person constitutes its "own" private time as the "absolute" time in relation to some common or absolute time. Dunbar's logic shows something completely different, namely, that "private" time constituted in this way is in no relation to any other time or to the time of the other, and that for this reason private time *is* absolute time.

This is the point of Berkeley's question: "Why time in pain, longer than time in pleasure?"[131] The mind that blindly accepts what hurts it has become pure pain and has to live the pain that it has become. It has to live as suffering, into which it sinks absolutely, without any reminders. The mind is the actual existence of the life of pain. It is the motion of pain, the time of pain. Outside of that time, there is no other time with which this "private" time could be in relation, a time that could promise the cessation of pain. The pain is absolute, and therefore, the time of pain is absolute time.

Berkeley here anticipates Bergson, for the whole truth of Berkeley's or Dunbar's logic resides in the insight that "private" time is lived time and therefore absolute time.

> If I want to mix a glass of sugar and water, I must, willy nilly, wait until the sugar melts. This little fact is big with meaning. For here the time I have to wait is not that mathematical time which would apply equally well to the entire history of the material world, even if that history were spread out instantaneously in space. It coincides with my impatience. . . . It is no longer something *thought*, it is something lived. It is no longer a relation, it is an absolute.[132]

Lived time is the blind motion of sensations that constitutes the very life of the person, and it is absolute because within the structure that is not the structure of the Other, there is nothing that could connect it with the time of the other.

But if time is the succession of sensations, then it has to follow that all intervals of that succession are, in fact, nonexistent, for the interval is precisely the absence of sensation and, therefore, something about which we do not have any idea. "Time therefore being nothing, abstracted from the succession of ideas in our minds, it follows that the duration of any finite spirit must be estimated by the number of ideas or actions succeeding each other in that same spirit or mind. Hence, it is a plain consequence that the soul always thinks."[133] Which means that perception always perceives and that it cannot perceive something about which we do not have any idea—an interval. It is possible that there is a time of an interval of perception, a minute gap within sensation without any sensation, which would be, presumably, the case with "the state" of death. What is more, such interruptions are continuous. "Men die, or are in [a] state of annihilation, oft in a day."[134] Discontinuity is continuous. Every sensation appears after an interval that separates it from another sensation. But since the mind does not perceive the interval, it does not know anything of it. It does not know death. It is always alive, even though it dies "oft in a day." The soul always feels.

However, even though time is the succession of sensations, of objects or locations, time is not space. Time differentiates itself from space, thanks to the motion of the volitions that moves sensations so as to constitute time. Both time and space are multiplicities, discontinuities of differences. But space is a discontinuous multiplicity of different locations, different topographies of sensations within a collection, whereas time is a discontinuous multiplicity of sensations moved by particular volitions. In the case of space, it is a matter of the multiplicity "of juxtaposition, of order." In the case of time, it is a matter of the multiplicity "of succession, of fusion, of organization, of heterogeneity, of qualitative discrimination."[135] In one case, it is a question of numerical multiplicity. In the other case, it is a question of a constellation in motion that cannot be juxtaposed according to the demands of simultaneity, but only according to the "logic" of succession that makes of it a qualitative multiplicity that, as the multiplicity of succession, can be only the continuity of discontinuity. The mind constantly thinks or feels, but that constancy means only that the mind constantly becomes some other body without knowing anything about the interval between those two "nows."

Needless to say, in the case of private time, "now" is not an untraceable and unthinkable moment that always already escapes in "not yet" or "not anymore." Rather, "now" can persist infinitely long, as long as there is the same constellation of sensations, until a "new" constellation of sensation eventuates, and until, in that way, the mind becomes another body, some other no one. Again: this is what Berkeley means when he says that man is in the state of death or complete annihilation very often during the day. The mind does not know anything of all of those innumerable deaths and identities. It feels only the body that it is at the moment at which it feels it.

Is this to say that the mind does not have a past, a memory that could preserve and save the past, or that the time of the mind is timelessness deprived of any "before" and "after"? No, for this discontinuous mind has a past and a memory, though it does not have a memory that would preserve the past as an "image" of what has passed. It does not have a memory that would nourish the "past" actuality by maintaining it in the form of an actual virtuality. In a sense, what has passed, the past tense, has passed completely. It has disappeared. It does not exist as an image that can be evoked. What has passed has been emitted into some other constellations. The past is the time that has become space, the time that disappeared, or in Proust's interpretation: "every moment of our life, as soon as it fades away, incarnates itself and hides in some material object."[136] The past that is now a thing will become time only in an encounter in which we encounter past time as a sensation, for "every moment of our life departs, as the soul in ancient beliefs, to incarnate in some object, in some particle of matter, and remains there imprisoned until we meet this object. Then it is released."[137] Then the mind becomes what it encountered: it "enters" into the body that is its past, becomes its own past, or the past itself becomes the "living" present, enlivened by the motion of sensations.

This time, therefore, has the structure of inverted time. Inverted time is not *chronos*, even though at first glance it may appear to be. Granted, *chronos* is nothing other than a "gigantic now." But the "now" of *chronos* is different from the "now" of inverted time. The chronological "now" exists by absorbing within itself the past as what "from the passion remained in the body." In other words, within chronological time, the past is what persists within a "now" as the motion of past passion that "from the depth" of the past can "subvert" the present, manifesting itself on the surface of the present. The past is the depth of the surface that is the present. Or, it is time that moves within another time and is enclosed in it. The past, therefore, is not the "order of juxtaposition." It is not the time that became space that will become time again. That is why to say that according to Berkeley, "every one is the standard of his own chronology"[138] misses the point. For according to Berkeley's conception of time, no one is his own chronology, because time is not chronological.

But neither is it the "time of the 'I'" (*Ich Zeit*), nor is it the time that is the very essence of the subject of geometrical perspective. In the structure of geometrical perspective, the point of the gaze "apprehends" itself only by projecting itself into the vanishing point, by a projection that escapes itself in the very moment in which it falls into itself: what returns to the gaze from the vanishing point is not that gaze itself, but another gaze, the gaze of the vanishing point. That is why the gaze cannot see itself. It is fundamentally blind to itself. The time of the "I" structures that "blindness," for the time of the "I" is time without the present. It is the time within which one leaps from the past into the future: from the not anymore, there immediately emerges the not yet. The "I" always moves toward its own not yet, toward the future, apprehending itself only as what it is not anymore, as the past, thus escaping itself, not managing to see itself. Its apprehension is always a misapprehension.

Inverted time reverses "perspectivist" time in the same way in which the subject of inverted perspective is the reverse of the subject of geometrical perspective. Here, within inverted time, nothing moves toward the not yet. On the contrary, everything moves toward the present, and it does not move as the past, as time hidden in the "depths" of the surface of the present, but as space, as the body that comes from the outside. In all the examples that Florensky gives in his analysis of inverted time, what moves toward the present remains "body," the body being either the cause that he designates as Ω, which is the external cause that from the regime of the "time of waking" transports itself into the regime of "dreamt time" ("dreamt time" is organized according to the logic of inverted time), or the event "X" that is the ploy of the whole drama of inverted time, its actual cause, but a cause that appears only after its consequence, which is the event "a." After the event "a," the event "X" moves toward its own consequence, toward the "a," as the movement of the past event, as the movement of the location or of the body that wants to meet its present. That is to say, the past of the event that resides in its future moves "now" toward its consequence in order to meet it within a "now" as a body, as a space that longs for its own time of appearance.

Space "desires" time. Or, to put it quite simply: the past is always the future, and therefore, the future of the past as present moves toward its presence, toward the event "a." Florensky uses the structural difference between the "time of waking" and "dreamt time" as an analogy for the difference between chronological and inverted time. If the time of waking moves from the past via the present toward the future, then inverted time moves toward the present in such a way that the past becomes the presence of the present. But the ruse of inverted time is that the "past" starts moving from the present, the cause causes within the present. The causal chain, the chain of succession, is inverted. Everything "begins" from the present. There is no past that precedes the present. The past cause of the present, the "past" event, moves toward the present as the cause caused by its effect. It is the cause that in the encounter with its effect falls into one with it and reshapes itself, transforming itself from the multiplicity of space into temporal multiplicity. In other words, it is always a matter of the "actual" event, "we repeat, we repeat, one and the same actual event is considered to be either the Ω or the 'X.'" It is always a matter of something that comes from the outside, from pure, immobile space. It is always a matter of the sensible idea that, through the sensible encounter, injects the motion of blind will into the multiplicity of space, makes of it temporality, the "now."

To say that everything moves toward the present means that a constellation of sensations is composed and encountered again, thus becoming the present: "we do not proceed from an actual present to the past . . . we do not recompose the past with various presents, but we place ourselves, directly, in the past itself. . . . This past does not represent something that has been, but simply something that is and that coexists with itself as present."[139] The person "falls" through itself into the outside, thus becoming the outside. The past is made present precisely by this falling, by this introduction of the order of space into the present motion of the will. Time "has inverted itself through itself."[140] The past coexists with itself as present, and the subject inverts itself by itself through itself, thus becoming the icon, as Florensky says, a verb in God's visual writing. The iconographic subject subjectivizes itself by falling into iconography, by circulating from one icon to another, from verb to verb, in accordance with the logic of inverted time.

EYE

Things

Covered things are staring at O. He can feel that he is perceived. Exhausted by percipii *he gives up his struggle for his own invisibility. He falls back in the rocking chair. "O is now seen to be fast asleep, his head sunk on his chest and his hands, fallen from the armrests, limply dangling." But precisely when everything is silent, when O has withdrawn deep into his interiority, "E resumes his cautious approach." The gaze of God penetrates O's interiority and wakes him. He is awakened by a word-image-body that opens his eye and enters it. Exteriority is now within his interiority. There is no distance anymore between the things and the eye. O sees: in front of him, in him, around him there is the gaze of E. Everything is the gaze of E. "E's gaze pierces the sleep, O starts awake, stares up at E. . . . O half starts from chair, then stiffens, staring up at E. Gradually that look." Gradually O enters the gaze of E and the gaze of E enters O's gaze. Everything merges into a single gaze. Everything becomes a single eye.*

Inverted Image and Specter

> Of course what most often manifests a look is the convergence of two ocular globes in my direction. But the look will be given just as well an occasion when there is a rustling of branches, or the sound of a footstep followed by silence, or the slight opening of a shutter or a light movement of a curtain. During an attack men who are crawling through the bush apprehend as a look to be avoided, not two eyes, but a white farmhouse which is outlined against the sky at the top of a little hill.[1]

These are the words of Sartre, but Sartre here speaks in the name of Defoe's Robinson, from his place, as it were, from the place that presupposes the world structured like Robinson's island: neither the world without others nor the world in which others are simply encountered, but the world in which all others are structured in accordance with the structure of the Other. It is a world in which a house is outlined against the sky in a single image. A house at the top of a hill is never "alone," even when there is nothing around it. This house is outlined against the sky. It is a house "with" the sky, a point in relation with another point, and therefore it is always already two points connected by a line. A single point is always already a map.

When it is said that the visual field is formed so "that the other is looking at me in every moment," when it is said, as Sartre has said, that this other does not necessarily have to be a subject, that it can also be an object (a house, a curtain, the branches) for which I am the object, then, of course, what is being posited is that things are looking at us. However, this assertion is possible only because any two points (the one comprising the house, and the other that is its reflection in the sky) are connected by lines that produce a geometrical projection of the perceptual field, a chart in which every point is an eye that manifests the possibility of a gaze. Men who during "an attack" crawl through the bush observe such a chart. From this chart, they "read" that in the very moment of this reading there are invisible points, and that each of these points is an eye that represents a gaze. The house is precisely an eye: "Now the bush, the farmhouse, are not the gaze: they only represent the eye, for the eye is not at first apprehended as a sensible organ of vision, but as the support for vision."[2] Each of these points is an eye in its double "nature": the place from which one can see and the object that might be seen. That is why the structure of the Other is not only this network of eyes capable of the gaze, but also the possibility for this entire network of eyes to be seen from one point, from the point of a single eye. For example: men crawling through the brush are looking at this map of eyes that do not yet see them. They are, therefore, those for whom "a white farmhouse is outlined against the sky," that is to say: they do not see only the spot from which they themselves might be seen, but also two other eyes (the house, the sky) looking at each other. To see the house outlined against the sky means to see a reflection or the image of an eye in another eye (the reflection of the sky on the surface of the house). To watch two eyes that are looking at one another is possible only from the position of a third eye. These men, hidden behind the bush, are in the place of the Third.

What is it, therefore, that this Third sees? He sees a curious thing: a "how." In other words, he sees what is unfolding between these two eyes, between an eye and its "outline" on the retina of another eye. And this "what" is nothing other than "how." For the third eye, the answer to the question "What do I see?" is always the answer to the question "How does one see?" Placed in the location of the Third, the "I" always sees how one sees. Which means: this "'I" sees how light rays are transferred from an object (an eye) to an eye (other object). It sees how these rays intersect in a single focal point at the retina of the eye and how this point forms the center of the retinal image that depicts "alive and intensive colors." However, when it "considers more carefully" what the eye it is looking at sees, the third eye also sees that "alive and intensive colors," reflected on the second eye, are extended in very regular geometrical forms, all of them rectangular. The third eye is surprised that the other eye sees what it sees—these colored rectangles. It is surprised because it had relied on a map that was already produced by some other third eye according to which it had expected that the forms seen by the other eye should be semicircular. After a moment of astonishment caused by the discovery that the map it holds is not correct, this "I" will make some changes to the existing map. It will make a new diagram and formulate a new law that reads as follows: As an effect of the way in which light rays are refracted, the retinal image of the eye will always depict rectangular colored forms.

In the case we are describing, the gaze of the third eye belongs to Isaac Newton. He is in the place of Sartre's man who carries out an attack. In front of himself, Newton has a map that Descartes had depicted. Needless to say, Newton carefully examines the map, trying to find out where the hill is, where the house is, and how it is outlined against the sky. All these investigations of the maps collected in the atlas called *Dioptrique* Newton carries out for one reason only: he wants to know *what* he himself sees, the same thing Descartes wanted to know. But Newton cannot see that "what," just as Descartes could not see it. That is why he is looking at the eye that looks at the other eye. He is following the motion of the gaze of that eye along the line, which he identifies as the line of the light ray, and he sees the gaze of this eye caught and reflected on the retina of the other eye. He sees what is reflected on the other eye. All the time looking at *how* an eye sees, Newton all of a sudden sees *what* the other eye sees. Thus he concludes with a description of what the retinal image in his own eye looks like.

Descartes did the same thing. He observed how in the eye of a dead ox, a retinal image is formed "in the perspective." He watched a tennis match and later claimed that the angles of light refraction are those at which the tennis ball bounces off the terrain. In contrast to Descartes, Newton does not watch oxen. In Newton's experiment, the one eye is a small opening in the blinds on the window of his room, the other eye is a wall opposite the window, and the line that connects these two points is a glass prism. Newton's experiment with a glass prism is supposed to explain what we see by explaining how we see, just as it is also supposed to elucidate "the phenomena of colours" and thus, indirectly, to explain the perception of the visual situation of the object.

> I procured me a Triangular glass-Prism, to try therewith the celebrated Phenomena of Colors. And in order thereto having darkened my chamber, and made a small hole in my window-shuts to let in a convenient quantity of the Suns light, I placed my Prism at this entrance that it might be thereby refracted to the opposite wall. It was at first a very pleasing divertisment to view the vivid and intense colors produced thereby, but after a while applying my self to consider them more circumspectly I became surprised to see them in an oblong form, which according to the received laws of Refraction, I expected should have been circular.[3]

These "received" laws of refraction are based on the "maps" drawn in *Dioptrique* and *Meteors*, which Descartes himself deduced, among other things, through reformulation of the law of refraction received from the "map" of Willebrord Snell. Newton is surprised that an error crept into those earlier projections: forms that the other eye sees are rectangular and not semicircular. That is why the map has to be changed.

But changing the map does not mean negating the map. On the contrary, everything proceeds according to the logic of the diagram established through the eye of the Third in the structure of the Other. However, this "third one" who is standing "aside" will not only reveal that forms of the visible, which are reflected on the retina, have to be rectangular, he will also discover various other features of the retinal image: the width of the angle of the refracted form, the influence of the nature of the glass or of the lens on the "angle of incidence" of light rays, and the situation or the position of the retinal image itself. Through projections of images of the one eye to the other eye, through the mediation of the prism put in front of the wall, which means establishing a situation analogous to that produced by the *camera obscura*, the eye of the Third will find out that images of objects on the retina *are* inverted. The image of the object that appears on the retina does not appear in the way we see it, but in the way we do not see it.

This is what Descartes, Malbranche, Molyneaux, and Newton all discovered. They were all in the position of the Third who sees how, on the wall-retina, an inverted image of the object-eye that this wall looks at is formed. In other words, in the eye of the other, they were able to see the "truth of vision"—what their own retinal image looks like. They were able to see what we do not see: that the retinal image is inverted. We never see the inverted world. We never see what is down as what is up. We never see what is reflected in the eye. We never see with the eye. Rather, the eye acts merely as a blind screen that emits the projection onto other screens, onto the screen of the pineal gland or onto the screen of the cerebral cortex. The story of those other screens is introduced in order to explain why we never see the inverted world. Guided by this, the story of screens told by Descartes represented a step forward in relation to what "purely" geometrical optics described. The story of an entire series of screens is what a philosopher of nature adds to geometrical optics and to what interests the "optician."

The scientist-optician is interested only in the eye. That was the case with Kepler. Kepler did not go "further" than the eye. He wanted to see only how the mechanism of the eye functions. He discovered that the picture (*pictura*) that is the image on the retinal screen is the inverted representation of *idola*, which is to say that the picture is the inverted image of the *imago rerum*, of the image of the world such as it is outside of the eye, in its purity and invisibility. But Kepler was not interested in how it could be that this reversed image was never seen. Although he understood that the invisibility of the inverted retinal image opened up a new problem within the theory of perception of the spatial situation of the object, he nevertheless did not offer a possible solution to this problem. He thought that it was beyond the limits of optics, and he wanted to stay within those limits, focused on the problem of the eye as apparatus:

> The study of optics so defined starts with the eye receiving the light and ceases with the formation of the picture on the retina. What happens before and after—how the picture so formed, upside down and reversed, was perceived by the observer—troubled Kepler but was of no concern to him. Neither the observer looking out into the world nor the process of perceiving the picture formed is Kepler's concern.[4]

As he himself said, Kepler leaves the problem of this "before" and "after" to the philosophers of nature, who, starting with Descartes, all claimed that this inverted image has to be inverted again, either on the screen of the cerebral cortex (Descartes's solution) or through the imperceptible intervention of God, who assumes the "form" of the unconscious judgment of the observer (Malebranche's solution). But disregarding the differences between the answers to the problem imposed by the inverted retinal image, all these "philosophers of nature" claimed that perception of the situation of objects is an effect of a two-level or even three-level process. What we see "immediately" is never what is represented and inscribed on the "retinal wall." It is always mediated by a geometrical projection of the retinal image that we never see, which is pure invisibility. And the mind's eye can see the noninverted image of the visible thanks only to the mediation of "natural" geometry. Relying on geometrical calculations that inscribed their labor in the very act of vision, and adopting the theory of transmissions, geometer-opticians could come to the conclusion that the gaze never sees what is emitted from the retinal screen and that, therefore, vision is never immediate. We never see "immediate objects of sight," which are the pictures depicted on the concave surface of the retina. We always see through the mediation of other projections: the eye remains blind, separated from the gaze.

However, Berkeley claims something completely different. He claims that it is erroneous to argue that the inverted retinal image is an immediate object of sight that escapes vision. The theory of inverted images is based on the presupposition that "the mind, perceiving an impulse of a ray of light on the upper part of the eye, considers this ray as coming in a direct line from the lower part of the object,"[5] thus creating an inverted image of the object, a reversed world. But this assumption is incorrect because it does not take sight into account. On the contrary, it announces that the proper object of vision is invisibility, when instead, it is a palpable figure, an object of touch.[6]

Who ever saw an inverted image of the object of vision? Who ever saw the intersections of straight lines and angles formed by these intersections? "I appeal to anyone's experience, whether he be conscious to himself that he thinks on the intersection made by radious pencils, or pursues the impulses they give in right lines, whenever he perceives by sight the position of any object?"[7] No one has such experience, and no one ever sees anything "inverted." We always see things the way we see them, and in whatever way we see them, they are never inverted. The inverted retinal image is invisible.

The inverted image is not, as Kepler would call it, the *pictura*: it cannot, says Berkeley, be determined as picture. The inverted image is precisely an image, which means a figure inscribed on the surface of the eye, and it is its "nature" as inscription that makes of it an immediate object of touch. The inverted retinal image does not belong to vision, mediated or not. It is an effect of geometrical projection. It resembles a geometrical body projected as a tangible image on the wall of a room. If inverted images are not visible, it is not because the gaze constitutively sees only "mediated" images. It is because they do not belong to the visible—they cannot be seen by any eye, and therefore neither can they be seen by the omniscient eye of a Third.

An eye that tries to see itself in the other eye will never see its own inverted image. In this case, the retina of the other eye functions as a mirror, and like any other mirror, it gives only noninverted images. The same holds for the eye of the Third. "Though if we suppose a third eye C from a due distance to behold the fund of A, then indeed the things projected thereon shall, to C, seem pictures or images in the same sense that those projected on B do to A."[8] One can never say that the "upper" parts of the image that we see are projected into the lower parts of retinal screen thus forming an inverted image.[9] Even the later invention of the ophthalmoscope, though it made relations between three eyes more complex, was unable to place the eye of the Third in a situation of omniscient eye: "With an ophthalmoscope C can look into A's eye and see various things, including, with one kind of adjustment, reflected images from A's lens and cornea. All this is produced by light which C's instrument shines into A's eye and poor A is seeing nothing but a glare of light."[10] The retinal images of the eye A at which the eye C looks are "images" like any other images. In order to see the inverted image, the eye would have to be able to see its own retinal images as inverted. It would, therefore, have to be able to see simultaneously two images, an inverted one and the uninverted one, which is impossible.

Geometrical optics had tried to make possible this impossibility by "tripling" the eye, by introducing the eye of the Third, which would be able to look at the same time both from its own place and from the place of the other. It turned out that the third eye got back only its "own" uninverted images, and that it is structurally impossible to occupy the position of the other that would remain the position of the other. Even with the help of the ophthalmoscope, the eye C cannot see what the eye A sees, but only what it itself sees: "The one important result that may follow these observations is that they may shake the faith of C in his diagram and make him understand that he is not omnivident any more than omniscient, and that diagrams may be just mixed metaphors."[11] In other words: there exists no eye of the Third, if the Third is the one who sees from the place of the other. The important insight gained from these observations is that if there is no eye of the Third, then, once again, there is no structure of the Other.

Of course, if the presupposition that immediate objects of sight are invisible is a mere artifact of geometrical optics, it is so only to the extent to which geometrical optics itself is an artifact of a theory of subjectivity that comprehends the subject as the process of appropriation of what it lacks—its own gaze. Geometrical optics attempts to recompense this deficit in the subject. The gaze of the subject is placed behind the screen of an always blind eye in order to enable that subject to attempt to escape the impossibility of seeing its own gaze: by looking at the eye of the other, the subject wants to see its own gaze, remaining nevertheless distanced from this other. The eye of the other here has the function of the concave lens that emits the image of the object or the image of the gaze of the subject into the gaze of the subject. And it is this screen that protects the subject from falling out through the retina of its own eye into pure externality. This screen, therefore, has the double function of the *spectacula*. On the one hand, like dark glasses (*spectacula*), it makes the gaze of the subject invisible to the other. On the other hand, it is a blind concave lens that reflects "images" from externality, thus representing the visible for the subject. But because the retinal images are invisible and because the subject, by the mediation of the *spectacula*, obtains the image of this invisibility, because it can see the image of the invisible, in this second function, *spectacula* have the character of *specula*.

Specula is the name for mirrors invented by Roger Bacon. On these mirrors one could have seen what was invisible, invisibility itself. *Specula* "produced" things that existed only on their screen. *Specula* created images that had no "being," pure appearances without being—specters. This was the strategy of geometrical optics. It produced the visible existence of something nonexistent.

This is completely incomprehensible to Berkeley. He claims that the image is an invisibility that is *imagined* as perceived, for in fact, as Berkeley's analysis of the experience of the eye of the Third had shown, it is not perceived at all. The eye of the Third only imagines that it sees what, in fact, it does not see. It sees the specter. "The specter is also, among other things, what one imagines, what one thinks one sees and which one projects—on an imaginary screen where there is nothing to see."[12] Berkeley cannot understand how the situation of objects can be judged by means of something that is invisible, as lines and their intersections or optical axes and angles: "And for the mind to judge of the situation of objects by those things without perceiving them, or to perceive them without knowing it, is equally beyond my comprehension."[13] Berkeley cannot comprehend that the situation of objects can be judged by means of the invisible and that on the basis of the invisible, one can assert the existence and the visibility of the invisible inverted retinal image. Those who claim this see ghosts.

Here Berkeley repeats the warning of Roger Bacon, who, in 1267, explained that those who look at the *specula* in order to see there how one sees "will run to the image and think the things are there when there is nothing but merely an apparition."[14] Geometrical optics, therefore, relies on apparitions, and precisely this ghostly world is a condition of possibility for the subjectivation of the subject, at least as far as Shakespeare and Descartes are concerned. What sets in motion all the procedures of Hamlet's subjectivation is an apparition that mediates him and subjectifies him. However, Hamlet knows that what "calls" him is a ghost and, what is more, by describing the apparitional nature of what he sees, he at the same time accurately describes the nature of the retinal image. When he says: "My father, methinks I see my father,"[15] he is in fact saying: "it seems to me, it appears to me," I imagine that I see something where there is nothing that could be seen. But such a fantasy is structurally necessary for the structure of subjectivity. It is what moves Hamlet and establishes him.

Descartes claims the same. Descartes's subject is subjectified by a phantasmatic contrivance of the evil demon. The meditative subject is subjectified by the visibility of an invisibility. Like Hamlet, the meditative subject is haunted by an apparition. Before turning to ontology and proving the existence of the good God who will "protect" the subject from all evil, modern thought will "produce" an apparition. Hauntology is the condition of possibility for modern ontology[16] and of geometrical optics.

The whole point of Berkeley's criticism is that modern philosophy, ontology as well as optics, is based on specters of what one cannot know and cannot see. The specter is precisely what one does not know as present, what is not present in its presence. The specter is not known, "not out of ignorance, but because this non-object, this non-present presence, this being-there of an absent, no longer belongs to knowledge,"[17] or never belonged to it, "at least to that which one thinks one knows by the name of knowledge." The specter is what is constitutively unknowable. It is something, some "thing," as Hamlet puts it, but that thing is precisely "nothing." (The difference between Hamlet and Descartes on the one hand and Berkeley on the other is that Hamlet thinks that this "nothing" is something, whereas Berkeley maintains that this "nothing" is nothing.) Geometrical optics thinks that even though "the Thing is still invisible,"[18] it is nevertheless something because it is seen as the thing of nothing.

Hamlet. The king is a thing—
Guildenstern. A thing, my lord?
Hamlet. Of nothing.

And it is precisely this "thing of nothing," which always "slips through your fingers,"[19] that is the condition of possibility for the subject and its gaze. Without the specter that casts doubt on hyperbolic excess, the *cogito* would not establish itself. Without spectral inverted images, the gaze of the subject would not be possible. Without this spectral image, the gaze would fall into the eye. The mind would become the eye, slide into the immediacy of the image, and become the object. However, Berkeley wants to examine the world without specters. He rejects the gaze produced by spectral images. Which is to say that he abandons the structure of the Other in which the Third is possible only on the condition of the spectral existence of the other. The Third is possible only if the other is a ghost and, conversely, if the visibility of invisibility is rejected, if the work of the specter is negated, the Third is impossible. And where the Third is impossible, the gaze is no longer distanced from the retinal image, and the retinal image is no longer inverted.

In fact, one cannot speak of retinal *images* anymore. There are only *pictures*, only immediate objects of sight. Those immediate objects are the objects seen by the gaze that is not distanced from the eye. Once again: immediate objects are pictures seen by an *innocent* eye, the eye of a child or an idiot, or, indeed, by the eye of anybody else, because everybody sees in the way children and idiots see. The innocent eye never sees inverted images, never sees some non-object: "To me it seems evident that crossing and tracing of the rays is never thought on by children, idiots, or in truth by any other, save only those who have applied themselves to the study of optics."[20] And those who applied themselves to proving how we are separated from externality by a screen of inverted images have devoted themselves to ghost hunting.

Of course, there are exceptions, even among them. There is, for example, a note by an oculist from Dresden who in 1583 explained how one sees much better without inserting the screen between the gaze and the visible: "It is much better [and more useful] that one leaves spectacles [alone]. For naturally a person sees and recognizes something better when he has nothing in front of his eyes than when he has something there."[21] Naturally, a person sees much better when *spectacula* are removed, when there is no retinal screen that separates the gaze from the visible. *Spectacula*, the spectacles about which the unknown oculist speaks, are both spectacles in the sense of glasses and spectacles in the sense of sights or of scenes. A spectacle is a vision that is distorted through the mediation of a screen. These spectacles produce both the representation and the deception, the representation and the nonvision, representation as nonvision. They produce specters. That is why it is better to give up spectacles and to see naturally, immediately, innocently, and idiotically. Idiots never look at the eye of the other. They are not subjects. The idiotic eye remains ignorant of the theory of transmission or hypothetical entities whose nature is spectral. It is ignorant of light rays and optical axes and angles. It sees only what it sees, and it sees only immediate objects of sight. The idiotic eye cannot answer the question of *how* it sees, it can only *describe* what it sees. This is Berkeley's fundamental idea—to free the eye and the gaze from all mediations and artificial structures, to come back to the innocent eye and to determine what the immediate objects of sight are. One does not look for the answer to the question "how" anymore. Now, when there is no eye of the Third, the only possible question is—*what?* What do we see? What are the proper objects of sight? All that is left is a lonely idiot with an innocent eye.

Gaze and Glance

The difference between Kepler's and Descartes's "theory" of subjectivity derives from the fact that Kepler was interested only in the eye. Kepler determines the subject by not determining it. For him, the eye is a camera or an apparatus, a "mechanism supplied with a lens with focusing properties,"[22] which is a determination that would be taken over by Descartes. However, while concentrating exclusively on analyzing the functioning of the apparatus, Kepler also speaks about the "deceits of vision," about the delusion of the gaze. The discovery of that delusion resulted from an attempt to explain the apparent changes in size of the diameter of the moon. No analysis of relations between heavenly bodies could offer the reason for these changes. Astronomy claimed that the magnitude of the diameter does not change, even though the eye sees the changes. To that, Kepler replied: if the eye nevertheless sees the change, then the reason for it would have to be in the eye itself. If the change in the size of the diameter is a delusion, then this delusion has to be the effect of the mechanism of the eye, whose lens, by means of focusing, distorts, lessens, or enlarges the diameter of the moon or the longitude of solar eclipses. The perception of the lunar diameter as smaller during a solar eclipse is due to a certain functioning of the optical mechanism itself.

The contours of a conception of the gaze and subjectivity are already clear. Kepler speaks only of an optical, desubjectified mechanism of the eye. But by not speaking about the observer and by referring to the deceptions of vision, Kepler is pointing to the nature of the gaze and its location. He is saying that the gaze cannot be identified with the eye. The eye is a mechanism that deceives the gaze and "distorts vision." What is distorted by the eye is nothing other than the image depicted on the apparatus, on its retina. The gaze, therefore, is nothing other than the image. This, of course, complicates the question of locating inverted images. If images on the retina are vision itself, and if we never see inverted images, then where are they? Kepler leaves the answer to this question to the philosophers of nature. What he does not want to leave to them is the insight that the image itself is vision. That the image is vision distorted by the eye means two things. First, there is no identity between the eye and the gaze. Second, the gaze is not placed behind the eye. The gaze is the image that is the effect of re-flexion of the visible in the eye. Or, the image is the re-flexion of the visible, deformed by the labor of the apparatus. The eye intervenes into the visible that is depicted on its retina. Only re-flexion that has suffered the intervention of the eye can be the image, and only this image can be vision. By placing the gaze within the eye, Kepler established a theory of subjectivity.

The fact that the gaze is exhausted in the image means that the subject is placed in the place of the image in the place of the gaze. That is why one cannot say that Kepler was not interested in analysis of the position of the observer. This position is determined by the identification of the image with the visible: *ut pictura, ita visio*. The observer or the subject is in the place of the gaze that is the image—the subject is the image. The subject is identified with the images of the world. Between subject and image there is no distance. Which is not to say that there is no distance between subject and externality. This subject that is the gaze that is the image does not sink into a visible object. On the contrary, it is distanced from externality, withdrawn from it. The gaze or the image is not the object of vision, a visibility that is outside of the eye. *Pictura* is not "the world outside of the eye." The gaze is the image that emerged by depicting the external world in the eye. It is a representation of the external. Like any other "artificial" image (*pictura*), this one is also painted by the brush: "The word that he [Kepler] chooses for what the painting does is *pencilli* —or little brushes— very small brushes."[23] The image is thus painted with "small brushes" that are moved by the *imago rerum*, by "the world outside of the eye." The image is the image of the outside world painted by it as its own representation.

Here we have a case of what Derrida calls the "visor effect." The outside world zooms in on the image that is the gaze. It paints itself as what it distances itself from and for which it remains invisible. The gaze sees only what there is to be seen—the image. It does not see what painted the image. The world outside of the eye remains invisible to the gaze. It is the invisibility that sees the gaze, because it painted it. This is the "visor effect: we do not see who is looking at us."[24] The visor effect will be "enforced" by an additional intervention of the optical apparatus that will "distort" the image, and this "distorted" representation of the world in the eye becomes the only thing that can be seen. It is placed at an insurmountable distance from "the world outside of the eye." The distinction between *imago rerum* (the image of the world outside of the eye) and *pictura* ("the image of the world caught up in the retinal screen") thus distances perception from what is perceived. Because the distance between *imago rerum* and *pictura* is insurmountable, because the subject is produced within the "internality" of the apparatus, "Kepler not only defines the picture on the retina as a representation but turns away from the actual world to the world 'painted' there."[25]

The problem of adequacy between image and world disappears as the effect of the insight that such adequacy is necessarily impossible, always already blocked by the structure of the optical apparatus. Of course, the problem of adequacy could appear in some other place—in the relation between image and observer. Kepler resolves this problem by identifying the image with the gaze. By "eliminating" the problem of adequacy, Kepler was able to turn to "internality": to the eye and the world "painted" there. If, after all these reductions, it is still possible to speak about the retinal screen, then this screen can only be a "passive" screen, a screen that is a mirror, a gaze that is an image. The image is thus understood as the subject's own image. "There is, admittedly, an ambiguity in the notion of the subject's 'own image'; it can refer either to an image *of* the subject or an image *belonging to* the subject."[26] However, in Kepler's case, there is no such ambiguity, since the image that belongs to the subject is also the image of the subject insofar as the subject is nothing other than the image. From this rudimentary theory of the apparatus, it will be possible to develop the hypothesis that the subject begins to exist by means of identification with the "signified image." This "signified" of the image, this "ideal point" that is the gaze, will become the point starting from which the image begins to make sense for the subject.

Descartes will maintain the distance between the image and "the world outside of the eye," but he will also establish the distance between the subject and its gaze or the point of its gaze. It is impossible to identify the subject with the image because this identification annihilates the subject. To say that the subject is the image does not amount to subjectivizing the image, but to objectivizing the subject. "The subject, in short, cannot be located or locate itself at the point of the gaze, since this point marks, on the contrary, its very annihilation."[27] This, as we have seen, is Descartes's insight. The subject is the effect of a double distance: of the distance between the image and the world and of the distance between the image and the gaze. The gaze has to be absent from the image, has to be what is invisible for the subject and what the subject is trying to appropriate. The subject is the effect of this insight, namely, that "there is a significant difference between what is impressed on retina and what we ourselves experience as vision."[28] Only through this fundamental split between perception and the experience of perception is the subject established as the subject.

There is no reason to maintain the subjectivity constructed in this way, however, precisely because it is constructed. This is Berkeley's claim. In both cases (in the case of Kepler, as well as in the case of Descartes), the subject is a mere contrivance insofar as it is a consequence of the presupposed existence of the invisible, of the specter. Needless to say, this presupposition is illusory, for what is visible cannot be made of the invisible. "That wch is visible cannot be made up of invisible things,"[29] or, in other words, something cannot be made of nothing. But the presupposition of the existence of the invisible is not the only error sustained by these theories. There is more to it. Both theories are nourished by the existence of a distance, and distance is also the existence of the nonexistent. It, too, is a ghost.

Like the inverted retinal image, a distance is also invisibility. The gaze never sees distance itself. Distance is not an immediate object of sight because it is not the "proper object of sight." At first sight, by questioning or negating the immediate visibility of distance, Berkeley refers merely to Molyneux's famous premise of the invisibility of distance: "For, *distance* of it self, is not to be perceived; for 'tis a line (or a length) presented to our eye with its end toward us, which must therefore be only a *point*, and that is *invisible*."[30] But the fact that the eye does not see the distance between itself and the object does not mean, as far as Molyneux is concerned, that such distance does not exist. His optics is geometrical. It presupposes a Third for whom distance exists: "Consider again the two trees. There is, we said, a gap between them; I see that there is a gap and there is a familiar device for *measuring* such visible gaps by the *breadth* of one's fingers held out at arm's length."[31] The existence of distance is affirmed by the eye of the Third, for which length is the width that can be measured by certain devices. Molyneux's solution, therefore, reads as follows: I cannot see the distance, but he can. He can measure it. Distance is metric distance. It is understood as a line, and every object on that line is a point. Distance is a "mathematical entity" made up of "mathematical points."

But distance does not have to be understood as a measurable aloofness between object and eye. It can also be understood as a mark of the externality of the object, as proof that objects are outside of the mind.[32] Speaking of distance, both Descartes and Molyneux tried to answer the following question: "How can the eye tell the amount of distance an object looks to be from us?"[33] But in Berkeley's consideration of distance, there is no eye of a Third that sees length as width. Berkeley, therefore, proceeds simply from the perspective of a helpless eye that does not rely on the help of a Third. Let us assume that this eye is looking at the already mentioned trees. In this case "I cannot in the same way 'see the gap' between either of the two trees and myself."[34] In fact, I cannot "see the gap" at all. I cannot see that anything is distanced. I cannot see that anything is outside of me.[35]

When Berkeley speaks about distance, he is trying to resolve the problem of "externality," of the existence of objects outside of the mind. He introduces the problem of "metrical distance" only in order to prove the identity of the point of the gaze and the subject. Berkeley almost quotes Molyneux's premise: "For distance being a line directed end-wise to the eye, it projects only one point in the fund of the eye, which point remains invariably the same, whether the distance be longer or shorter," which, in other words, means that "distance, of itself and immediately, cannot be seen."[36] But since it is "quoted" in a different context, Molyneux's premise produces different consequences. It cannot be understood as an explanation of the way in which we measure distance, because there is no Third, no width, and therefore no length, either. Lines and points vanish. The mathematical entity becomes a sensible entity. All of a sudden, mathematical points become sensible points. Distance becomes a mark of the existence of things outside of the eye, but a mark that is missing.

That distance is "in its own nature imperceptible"[37] does not mean anymore that we will have to rely on someone else to measure the distance. It means that we cannot measure distance at all, because there is no such thing as distance. The eye is unaware of distance. For it, everything is absolutely close because everything is in it. That everything is in the eye is the effect of the nonexistence of distance, and distance vanishes because it is perceived only "as a point in the eye." The visible world enters the eye in an absolute intimacy with the eye. This immediate object of sight, this visible that is in the eye, is a picture. But the picture is the visible object itself, and not its representation. The picture is, therefore, a simulacrum that becomes one with the point of the gaze. Distances do not exist anymore: neither between gaze and image nor between the image and the object of vision. The picture is the gaze that is the image that is the object of sight. Once again, the subject has disappeared into the visible. And once again: where there are no distances, where the subject can become what it sees, there exists only an iconographic subject and its innocent eye. The gaze is inseparable from the picture in precisely the same way as in the primitive, "original" gaze that watches from an inverted perspective: "In primitive art the artist apparently considers himself (his visual position) as an integral part of the picture, seemingly forming a unified whole with it."[38] This experience of the totality of the eye and the picture is (again) the experience of the innocent eye: of the eye of a child, of the eye of a savage, and of the eye of a blind person who has just started to see.

CHILD

There is a supposition that the eyes of a child have an inverted perspective. This is what Berkeley supposes. He says, "suppose inverting perspectives bound to ye eyes of a child"[39] —suppose, in other words, that to the eye of a child, the image does not appear as a frightening otherness. This child would not have dreams in which it tries to get closer to something that always remains at a distance. The eye of a child annuls the Keplerian distance between image and world. Its gaze is identical with the image, and the image is "the world," a visible object. The child-observer is within the picture. Its gaze is upon the point that is the "vantage point," which is within the picture itself. And because the observer sees the image from within it, the proportions of the image get reduced, the background "getting closer to the foreground." However, the subject to whom the picture is given in the central perspective does not have the experience of a vanishing background. This subject is outside of the picture (of the visual field, of the painting). The picture is distanced from him, and because it is distanced, because the vantage point is the point at which "lengths" intersect, the picture also has depth. That is why proportions of the image can be shortened only for the eye that is inside depth itself, in the middle of the internality of the picture: "This phenomenon may be understood by suggesting that the diminution in size of the objects in the representation depending upon their distance is presented in the system of inverted perspective not from our viewpoint (the viewpoint of the spectator outside the picture), but from the viewpoint of our *vis-à-vis*, of an abstract internal observer."[40] But the "internal observer" is an abstraction only from the perspective of "central perspective." On the other hand, the eyes of the child are in the position of an internal spectator. Their perspective is inverted, which means that they are placed in the "point of distance," in the depth of the painting. And since the position of the spectator is at the point of distance, distance disappears—there is no depth anymore. That the "proportions of the image" are shortened for this eye means that every depth is annulled, that the difference between foreground and background has disappeared, that depth has become surface, and that the third dimension of the image has been lost. Not the internal spectator, but the third dimension becomes a construct, artificiality, and abstraction.

The eye of a child does not see "deep" images. "Depth" is not an immediate object of sight. The eye of a child that has inverted perspective sees only flat, "two-dimensional" images. Or, since that eye is in the core of the two-dimensional image, one could say that the gaze appears where the gaze is. The gaze is the object of sight—the picture. Children's drawings bear witness to this: "Indeed, children often interpret such forms in their own work with reference to an internal observer whose position is opposite that of a viewer of the picture.... The artist places himself in the position of the depicted figures and looks at space through their eyes."[41] The child-painter becomes each of these painted figures or faces. The faces fall into his eye, and their eyes become his eyes. But if the image is in the eye, if the world is in the eye, where, then, is the eye itself? Is the eye within the picture, or is the picture within the eye? However, this question is senseless, because identification of the gaze with the image, of the subject with the object, cancels the distinction between outside and inside: outside is inside insofar as this inside is outside. Yet again: outside of the mind thus means inside the mind, and vice versa. That there is nothing outside of the mind means precisely that the mind is what is outside of the mind.

SAVAGE

Another case of the innocent eye, another eye that does not recognize representation, the eye for which nothing is deep: the eye of a savage also posits itself into the picture that is both object and gaze. This eye does not recognize projections. If a savage knows of the map at all, then this map is the impossibility of projection: "in primitive maps ... lines indicate a general inability to draw."[42] If the basic function of the map is to make visible "what is otherwise invisible,"[43] then in the case of "primitive" maps, the map loses its function. Such a function for the map is impossible here because there is nothing that is invisible for the innocent eye. No representation exists to cover the visible. The map of a savage is not a representation. It is a mad map, an annihilation of the distance between the gaze and the image that is the visible object. The map is at the same time the eye and the visible object. The map of a savage is the eye becoming the image and the image becoming an eye—the picture. Such a savage map is "formed" as an eye, surrounded by chaotic "uncartographic lines." Those chaotic lines are the pictures of what is "unpredictable," of what emerges suddenly, as an event, without announcing itself on various horizons or in the depth of their intersections. These lines are the picture of unpredictability and the absence of any causal relation between visible objects. The map of a savage is simply an eye or a picture with no horizon: "It is interesting in primitive maps one often observes a net of haphazard, tangled lines having no relation to the map."[44] In the world of an innocent eye, there is no causal connection between two visible points.

THE BLIND PERSON WHO HAS JUST STARTED TO SEE

"Molyneux's man." Molyneux imagined a person born blind, a person who knows the world only by touch, to whom sight is given back and who all of a sudden is able to see the world for the first time with a supposedly innocent gaze.[45] Molyneux's question was: What is it that this person would see? What does the world look like when seen by an innocent eye? This "What does the world look like?" implied an answer to several questions: whether the person who has just started to see would be able to *see* the difference between the forms of objects that it recognizes by touch, whether it would be able to name these objects,[46] whether it would see distance, and whether it would see depth.

Different answers to these questions were offered.[47] Berkeley, as we have already mentioned, claimed that the innocent eye does not see distance because distance is not the proper object of sight. In 1728, a surgeon named Cheselden reported to the Royal Society that he had performed a cataract operation on a patient born blind. Cheselden's report on what that patient saw, his report on what the first, innocent gaze looked like, confirmed Berkeley's analysis of an innocent eye. Cheselden's patient had not seen distance, nor had he seen anything at a distance from himself. "When he first saw, he was so far from making judgement about distances, that he thought all objects whatever touch'd his eyes (as he express'd it)."[48] This innocent eye saw the objects within itself. It saw everything as if everything were within itself. The visible object, which is in the eye, constitutes the totality of the visual field. It exhausts the visual field in which, therefore, there is no amorphous "space" of surrounding objects. In other words, this eye does not know of the horizon either: a man who has just started to see is unable "to imagine any lines beyond the bounds he saw."[49] This is precisely what Berkeley had claimed less than twenty years earlier. He had claimed that for an innocent eye, distance is imperceptible, that it is not the proper object of sight. "Blind at 1st could not know distance,"[50] "all things to a Blind at 1st seen in a point,"[51] and this means that things are in the eye—it sees them that way.

But the fact that it sees things within itself is not the complete answer to the question regarding what the image that is not distanced looks like. The experience of a child's innocent eye that sees by means of an inverted perspective could offer the answer to this question: to the eye that is "in the depth" of a painting, the proportions of that painting are reduced, foreground and background coincide, the depth bursts upon the surface, and the image becomes two-dimensional. This is what "Molyneux's man" sees. He sees images without depth, flat images.

However, under certain conditions and in certain circumstances, the absence of distance does not have to suggest the absence of depth: "a picture of, say, just a globe seen through a stereoscope will look bulgy or in depth but, without any other distance cues, it won't look to be at any particular distance."[52] But this differentiation between seeing distance and seeing depth misses Berkeley's point. Even if we leave aside the fact that in this case, the image is produced by the mediation of the "stereoscopic eye" and that therefore it cannot coincide with the experience of the innocent eye, even then, this differentiation cannot be maintained. A visible distance would have to imply the possibility of seeing "mediate ideas," of seeing the juxtaposed points, or of seeing the images that are alongside the line of the gaze. It would imply that the object of vision is never in the foreground, or, differently, that it is the foreground that is the background of a foreground that remains invisible. In other words, the play of foreground and background implies that the perceptibility of distance is at the same time the perceptibility of depth.

According to Berkeley, this is impossible, for it presupposes the existence of "mediated ideas," which means either between my eye and the distant object there is nothing (in which case it is not a distant object), or between my eye and the object there are some "visible points" that would have to remain invisible in order for my eye to see something as distant. This is also impossible, since what is visible cannot be invisible. Therefore, those "visible points" or "mediate ideas" have to be visible. In this case, they are "hiding" the distant object, and my eye sees only the closest of those "ideas." And the closest idea is the one that is within the eye. That is why there is no distance, and that is why imperceptibility of distance must be understood as imperceptibility of depth.

The eye that "believes" that it perceives depth or distance is either deluded or its perception is not the effect of visual perception at all: "Likewise, sometimes it seems that the bulge on the foreground of a tomato can be immediately seen. It seems that the depth is immediately seen. Now, if we understood Berkeley well, he claims that whatever comprehensions of common sense might be, I can never immediately see the bulge on the face of a tomato, that is, the depth or the distance is never immediately seen."[53] Properly speaking, depth and distance are not the proper experience of vision.

> Berkeley's argument, made quite explicit, runs roughly like this. What I call depth is in reality a juxtaposition of points, making it comparable to breadth. I am simply badly placed to see it. I should see it if I were in the position of a spectator looking on from the side, who can take in at a glance the series of objects spread out in front of me, whereas for me they conceal each other—or see

the distance from my body to the first object, whereas for me this distance is compressed into a point.[54]

For me, for my eye, therefore, nothing is ever distant.

At first sight, everything unfolds the way Locke described it when he tried to discover how an innocent eye sees and what is seen "immediately." Locke does not doubt that the eye of a child sees flat, two-dimensional planes: "When we set before our eyes a round globe of any uniform colour, v.g. gold, alabaster or jet, it is certain that the idea thereby imprinted in our mind is that of a circle, variously shadowed, with several degrees of light and brightness coming to our eyes."[55] For Locke, however, that distance and depth are imperceptible does not mean denying the externality of the visible object, since the idea of externality is the effect of "our judgment." It is an act of reason that, on the basis of a series of different experiences formed into a habit, asserts the externality of the visible. In this way, in every act of vision there is an action of "judgment" that "presently, by an habitual custom, alters the appearances into their causes. So that from that which truly is variety of shadow or colour, collecting the figure, it makes it pass for a mark of figure, and frames to itself the perception of a convex figure and an uniform colour, when the idea we receive from thence is only a plane variously coloured, as is evident in painting."[56]

On the basis of the same presuppositions, Locke draws conclusions different from Berkeley's. Locke claims that although the eye, "in itself," sees everything two-dimensionally and not as distanced, it nevertheless always sees everything three-dimensionally and as distanced. He claims that although for the eye nothing is convex, everything is nevertheless seen as convex, "as is evident in painting." Locke can claim this because, like Descartes, he maintained the distance between subject and visible and places the gaze beyond the flat surface of the retinal screen, which, because it is itself flat, deforms three-dimensional images into two-dimensional images. This is Locke's explanation of the existence of two-dimensionality: the retinal screen, like a painting canvas, is two-dimensional. All images that are painted on it have themselves to be two-dimensional: "Since the first inlet of the markings is through senses, they are inscribed in a flat surface.... Thus since the [retinal] wall is flat, so are the pictures."[57] In other words, the "spatial characteristics" of the visible that is painted on the retina have to be in accordance with the "spatial characteristics" of the retinal "wall"—the "constancy hypothesis." However, according to Locke's explanation, the constancy hypothesis is subverted precisely by the nature of the retina, and our reason has to reestablish it. Reasoning understands the two-dimensionality of the image as a necessary effect of the nature of the plane of the retinal wall. The two-dimensionality of the image is a deviation from the three-dimensional visible, a deviant representation of the visible caused by the structure of the optical apparatus. That is why reasoning has to overcome this deviation and to affirm the constancy hypothesis. That is why reasoning reads the two-dimensional image as a sign of the three-dimensional image and the absence of distance as a sign of its presence.

But for Berkeley, two-dimensionality cannot be a deviation because every image is absolutely true. It is the object itself, the picture that is the eye. That is why we always see things from an "absolute closeness" that designates the disappearance of depth and distance. Everything is close, and everything is mere surface. This means everything that is visible "is given" as a close-up, if that close-up is understood as a oneness of the gaze and the image: "The close-up does not tear away its object from a set of which it would form part, of which it would be a part, but on the contrary *it abstracts it from all spatio-temporal coordinates*, that is to say it raises it to the state of Entity."[58] The close-up as an entity is a sign of a "fervor" that draws the body out of itself and draws the mind out of itself. The circle is closed: there is a fervor once more. But, at the same time, the circle is opened, for this Berkeleian fervor is not the fervor of totalization, not a manneristic frenzy. Rather, it is a fervor by means of which the eye annihilates the distance and becomes the picture, an object among objects, a part of the totality. The anonymity of the gaze of the iconographic subject is the effect of this partiality or of this incompleteness: "sensation can be anonymous only because it is incomplete."[59] There is no pretension to totalization anymore.

The eye becomes a visible object always "given" in a close-up, an "entity" that exists "in itself," a simulacrum, an idea. But ideas, precisely because they are given as a close-up, without distance, are also given without depth. They become mere planes. They create a plane universe, the iconographic "wall" on which the Berkeleian-Beckettian anonymous person becomes a flat spot. This is the effect of the "logic" of the close-up: everything unfolds as if in some point, in a "point" of a close-up, the entire depth of this imaginary space were to emerge on the surface. The close-up tends to annihilate the sharpness of depth, and with it, perspective. It is something like a "magnified" and "inserted" point that becomes a flat, visible object. Every idea-simulacrum is thus at the same time a close-up and an insert, an absolute closeness and an object "for" itself, an object that is not connected with other objects within some presupposed or imagined visual field. Ideas, simulacra, subject-objects, pictures—all are flat surfaces. They are pure percepts, if percepts are not merely an effect of perception but "*perception in becoming*," not the perception of becoming, but the perception that becomes what is perceived. Perception in becoming is an "absolute potentialization of the gaze" that is possible only as an effect of an innocence by which the gaze becomes the picture,[60] or the subject, or the object.

But, if everything visible is a close-up, then everything visible is a face. "The close-up is the face."[61] However, the face that appears in Berkeley's optics is not the face that appeared in Descartes's optics. It is a matter of two differently structured faces. In Descartes, the face functioned as a substantial support for the maps, as a papyrus in which different geometrical projections were inscribed, as a screen on which the form of the visible was projected. The visible projected on this face was always projected as an image, as Berkeley would say, which means not as an immediate object of vision, but as a "*figure.*" In Descartes's optics, projection always projected another form, and all of those forms were nothing other than different "*expressions*" of the one and the same face—the substance. The face that projects an image is a reflexive face.

But there are differences among faces. In the case of Berkeleian optics, the face is not an "immobile reflecting surface." It does not function as substantial ground upon which forms and figures are engraved. Here, the face is what is engraved, what is "substantially" changed with every new picture—the face is the picture. This face is an "intensive" face. "Here the intensive series discloses its function, which is to pass from one quality to another, to emerge on to a new quality. To produce a new quality, to carry out a qualitative leap."[62]

The intensive face is a series of different and unconnected faces, a "leap" from one face into another face, a becoming of a face. This "face" is an endless set of faces. It is a faceless face, an anonymous subjectivity without substance. But for that reason, the difference between the reflexive and the intensive face could be summed up in the following way: the intensive face does not have a steady form. There is no such thing as an outline of the intensive face. "Where, therefore, is the criterion of distinction? In fact, we find ourselves before an intensive face each time that the traits *break free from the outline,* begin to work on their own account."[63] The intensive face does not have form. It is an unformed surface: a face without a face. That the subject plunges into the visible object and becomes the picture does not mean, therefore, that "he" assumes a form, for the form is not the proper object of sight at all.

Now the complete answer to Molyneux's question is given. Let us repeat the question:

> Suppose a man born blind, and now adult, and taught by his touch to distinguish between a cube and a sphere of the same metal, and nighly of the same bigness, so as to tell, when he felt one and the other, which is the cube, which the sphere. Suppose that the cube and sphere placed on a table, and the blind

man to be made to see: *quaere*, whether by his sight, before he touched them, he could now distinguish and tell which is the globe, which the cube?[64]

This question does not ask whether an innocent gaze sees the difference in magnitude or in material, since the sphere and the cube are of the same material and of the same size. It asks whether the gaze can discern the difference between the forms or the shapes. More succinctly formulated, this question asks whether an innocent eye sees the shapes at all, whether form is an immediate object of sight. "No, form is not the proper object of sight," would be Berkeley's answer.

Form or shape is as imperceptible as distance and depth. An image, which is the proper object of sight, does not have shape or form. Cheselden's patient will confirm this. After starting to see, this patient "could form no judgement of their shape . . . he knew not the shape of any thing, nor any one thing from another, however different in shape, or magnitude."[65] Shape and form, figure and extension, which in order to be perceived at all have to have form, are not immediate objects of sight. The only immediate objects of sight are colors and light, nothing else. "W^t can we see beside colours? W^t I see is onely variety of colours & light."[66] The eye sees only colors, and colors are within the eye. They are the eye—they are the mind: "the colours, which are the proper and immediate object of sight, are not without the mind."[67] The eye that is the gaze that is the mind is nothing other than the picture, which is a bundle of colors.

That is why such pictures are metapictures. But metapictures are not autoreferential pictures that are "split" in themselves and that through themselves refer to themselves as to their own referent. The "structure" of reflexivity is not inscribed in metapictures. Rather, they are the pictures that are "beyond" any figure and any image insofar as the image is what represents the picture for the eye. Metapictures are pure presentations of the visible, the visible itself that enters the eye. "Perhaps the most obvious thing called into question by this metapicture is the structure of 'inside and outside.'" Their "form constructs an inside-outside structure that is continuous, without breaks or demarcations or duplications. It is a meta-picture in a strict or a formal sense . . . one that dissolves the boundary between inside or outside."[68] Metapictures cannot be autoreferential because they do not recognize demarcations. Their nature is therefore multistable. A multistable picture is a picture that "varies" and "vibrates" within itself. It is neither autoreferential nor referential because the referent is destabilized and negated by the instability of the picture; the "multi-stable picture annihilates auto-referentiality."[69] It is a constant mutation of stability.

Proper objects of sight, which are multistable metapictures, planes of variable colors, thus appear as "pure optical pictures," as "instants." In the visual world of the Berkeleian type, it is not possible to distinguish between two aspects of the optical picture, not possible to maintain the existence of both "constats" and "instats" insofar as the "former (constats) [give] a vision with depth, of a distance, tending towards abstraction, the other (instats) a close, flat-on vision inducing involvement" of the observer in the picture.[70] In the Berkeleian world, constats do not exist. There are no optical pictures that can "give" depth (there are no "reports"). There are only instats, planes of colors inducing the "involvement" of the subject in the image, inducing indiscernibility between subject and object. "As for the distinction between subjective and objective, it also tends to lose its importance.... We run in fact into a principle of indeterminability, of indiscernibility; we no longer know what is imaginary or real, physical or mental, in the situation, not because they are confused, but because we do not have to know and there is no longer even a place from which to ask."[71]

The knowledge of what is real and what is imaginary is now constitutively impossible because there is no longer any fundamental condition of possibility for it: a place from which to ask, the place of the subject. To see no longer means to see from somewhere and to see something that is "there." The condition that prescribed that the object could become visible only if the objects surrounding it formed its horizon is no longer valid. Vision is no longer "an act with two facets" in which "the inner horizon of an object cannot become an object without the surrounding objects' becoming a horizon."[72] There is no horizon that "guarantees the identity of the object."[73] The horizon guarantees the identity of the object by being its own periphery inscribed upon the center. "I direct my gaze upon a sector of the landscape, which comes to life and is disclosed, while the other objects recede into the periphery and become dormant, while, however, not ceasing to be there. Now, with them, I have at my disposal their horizons, in which there is implied, as a marginal view, the object on which my eyes at present fall."[74] However, the paradox of the relation between fixed object and horizon or between center and periphery is that the periphery can "belong to" a center only if the center is outside of it: the center is a center of some periphery only if it is outside of the periphery. Therefore the distinction between the object of vision and its horizons is "reflexive." It implies distance. When there is no distance, there are no horizons, and nothing ensures the identity of the object-subject. There are only planes of colors without any "margin." The marginal visible has withdrawn and is lost.[75]

This is the existence of the structure of the night. Distinct objects and a fixed point of view disappear because the "night is not an object before me, it enwraps me and infiltrates through all my senses, stifling my recollections and almost destroying my personal identity."[76] Personal identity is destroyed because all of a sudden, as an event, as an encounter, an object bursts out from the "heart of nocturnal space"[77] and hits me, or what I am, whatever I am, thus forcing me "to become united with it." The destruction of personal identity and the establishing of an innocent eye is possible only in the structure of the night, because only in the night am I "no longer withdrawn into my perceptual lookout from which I watch the outlines of objects moving by at a distance. Night has no outlines; it is itself in contact with me,"[78] as in some mad, pure optical tactility where every distance is annihilated. When, as Berkeley demonstrates, depth "quite obviously belongs to the perspective and not to the things," and "depth is only a moment in arriving at a perceptual faith in one single thing,"[79] then we are "called to find again the original experience from which the world originates," the world of pure colors bursting out of the night, entering an "I" that has become pure openness. The eye is no longer a window; it is an openness through which externality breaks in. The structure of the night is the structure of openness.

But when the window is open, and things crawl into the room, or when the "room" leaves the "room," what actually happens to the eye? What happens to iconographic subject? The iconographic subject falls into the icon, into the visual writing of God. Two gazes do in fact merge, but asymmetrically. Even though, through this merging, the gaze has become an icon, it has not become iconography, has not become the infinite gaze of God. It has become a word-image-idea-sensation in God's visual writing.

Everything unfolds the way Beckett had understood it. O fails in his effort to stay "beyond the angle of immunity." That is to say, even though he tries by every means not to enter the *percipi* (by covering externality [the Cartesian strategy], by covering the mirror, the image of God [the icon], and even the map) he is, nevertheless, perceived. "Covered" visibility is not invisible. It is visible precisely as the covered visibility that looks at O. So that even when O closes his eye, the gaze of E appears, crawls into O's "closed" eye and *opens* it:

> Halt and brief image, not far short of full-face, of O still fast asleep. E advances last few yards along tattered wall and halts directly in front of O. Long image of O, full-face, against ground of headrest, sleeping. E's gaze pierces the sleep, O starts awake, stares up at E. Patch over O's left eye now seen for the first time. Rock revived by start, stilled at once by foot to ground. Hand clutches armrests. O half starts from chair, then stiffens, staring up at E. Gradually that look. Cut to E, of whom this very first image (face only, against ground of tattered wall). It is O's face (with patch) but with very different expression, impossible to describe, neither severity nor benignity, but rather acute *intentness*.[80]

Gradually that look pierces the gaze of O until O opens his eye and sees a huge face, the close-up that is becoming closer . . . and then a cut, an interval, a pause, a break, and the face of O is all of a sudden the face of E, just a flat face on the wall, "acute intentness." O has not managed to escape the angle of immunity. Innocence does not know immunity. O becomes the picture on the two-dimensional wall—the iconographic subject becomes the icon in the iconography. The iconographic subject falls into the gaze of God and in this way becomes imperceptible. It has entered the field beyond the angle of immunity, thus becoming the gaze of E, thus becoming absolutely immune to the gaze of E. The subject disappears into the face on the wall that does not see itself. "Is this not precisely what is needed, to cause to be in order to become imperceptible, according to the conditions set forth by Bishop Berkeley?"[81] To be, therefore, means to be perceived, but only because to be perceived means to be unperceived.

But the gaze that is innocent cannot be called a gaze. Only the subject possesses a gaze, and he possesses it precisely because he does not posses it and he does not possess it insofar as he does not see it. The gaze is a blind spot inscribed upon the visible. The subject becomes the subject through an effort to appropriate what he does not have by an appropriation that mediates the blind spot of the gaze. The gaze is always an effect of self-mediation. It is an effect (or the cause, it comes down to the same thing) of self-reflexive thought. The gaze is nothing other than a mediated look. It is a look that returns to itself from the vantage point in which the subject cannot locate himself. It is a look that comes to itself from that vantage point in the process of the subject's "self-appropriation," in a process through which the subject focuses himself in the "point of the gaze" and thus transforms himself into a pure attention toward himself, into an attention that carefully observes and inspects "itself."

And it is through this self-establishing that the subject establishes itself as an "intelligibility" that is the condition of possibility for every intelligible structure, that is already inscribed in the experience of elucidation of the object: "Since in attention I experience an elucidation *of* the object, the perceived object must already contain the intelligible structure which it reveals."[82] The structure of the gaze is the structure of this reflexive circulation through which the gaze separates itself from itself and returns to itself, thus establishing the subject that will establish the object, and so on to infinity. The gaze is a guarantor of self-doubling and of the self-establishing of subjectivity. The gaze "does not refer primarily to looking but to expecting, care, attention, considering and guarantee, which is made emphatic by a prefix that refers to doubling."[83]

But this circulation takes time, which is to say that the gaze cannot be exhausted in a "now." The gaze lasts. It has a "story," its "before" and its "after." The gaze is historical. It "involves persistency, as if it is animated by a hope that it will add to its discovery or its conquering what escapes it."[84] The gaze aims at what is hidden—at the point of the gaze. But in order to discover itself and to return to itself, the gaze has to stabilize itself in a fixed point. That is to say, it has to sacrifice what is immediately visible: unstable planes of colors. The gaze "is ready to give up its power of immediate perception in exchange for the gift of permanent fixing of everything that escapes its grasp."[85] It will, therefore, have to structure the visual field. It will have to establish the relations between visible objects, thus, by the same token, establishing horizons. The gaze, therefore, is not only the power that "gathers the image" into a unity, the gaze is the power that *structures* the visible as such: "For, in this context, by the term gaze I do not refer that much to the ability to gather images, as to the ability to gather relations,"[86] the ability to establish the structure of horizons. The gaze connects unconnected and unstable pictures by "lines and their intersections." It brings measure and order into the wild flux of colors and lights. It disciplines the immediately visible by covering it with the sheet of the geometrical map.

Of course, in a world that does not recognize distance, the gaze is not possible. Where there are only immediate objects of sight, the flow of colors and light, there can exist only a look that becomes what it looks at. But this look is not vision. It does not see through "the hidden," because it never distances itself and never focuses itself. Therefore, where the "subject" is the picture, there can exist only the glance, rather than the gaze. The glance is a distracted, inattentive look, or the look "whose attention is always elsewhere."[87] The glance slides through the visible and merges with it without recognizing any distance or doublings. It is a "forgetting" of the gaze, an endless succession of pictures on the wall of iconography. It is always another picture. It is therefore without history. The glance has no past. It does not have its own aorist. It is a pure sliding of icons along the two-dimensional surface, a sliding that slides as in a kind of vertiginous or delirious perception.[88] As unstable visible surface, the glance does not close the visual field. It does not cover it with a network of intersecting lines, does not discipline the visible by imposing measure and order. The glance is openness toward an open visual space. It is a "categorial scandal in which . . . divisions . . . elide."[89] The glance always functions where the visible opens itself in an endless multiplicity of directions, to an openness in which there exists only a "yes-body" that could wish the "closeness to disappear," the "world to open," and every "here to become somewhere else," so that it, this yes-body can be peaceful in the outside world—in colors.

Optic and Haptic

But if immediate objects of sight do not have spatial features,[90] if in the immediately visible world there is neither distance nor figure, then this visual book of God manifests its paradoxical nature—paradoxical insofar as it is still the book, but the book that resolves the opposition, immanent to every book, between the book and its site. By subverting this opposition, the divine visual book also subverts that between the visible and the readable.

> Visible and readable are also opposed in another way, in the same way that site and book are opposed. The sense of place, its "genius" is itself local: it requires the being-there of colors, values and lines, and they are to be seen if what it is to be known . . . . The book, on the contrary, belongs to no place, its signification is emancipated from the site. . . . Thus, the man of the book has no land.[91]

In other words, the nature of every book is twofold: it is sensible-supersensible, both the visibility of the external and the readability (which is possible only from an internality) of what is external. However, the immediate objects of sight that constitute the divine book negate this dual nature of the book. In the divine book, nonsensible becomes sensible, and the visible becomes readable. The visible becomes the very sense of the readable because the readable is no longer "the object that is missing in the visible," but the object itself revealed in its presence—the visible. Through this negation of the dual nature of the book, the divine visual book appears as what "has no place" and what does not occupy space because it has no spatial features.

The reader of this book, who becomes one of its words, thus becomes the reader without land. He "enters" the book that is without place and travels through it, this displaced book, from word to word, sliding along the flat visibility of the readable, because images-ideas are not in the place of anything. They do not represent anything, they simply are. This means that "they function through their material and their *organization* . . . they conceal no content, no . . . secret of the work," because the force of the work "lies entirely in its surface. There is only surface."[92] To refer to the organization of the visible is to say that proper objects of sight are an organization of "*minimal visible elements*," an organization that establishes the proper object of sight. Or, in other words, it means that even though planes of colors do not have fixed form (in this respect Berkeley is explicit), they are nevertheless extended. But this extension is not spatial: it is the extension of the picture, and not of the image, a "strange" extension in which the figure cannot be distinguished, an extension that is rhythmic, turbulent, and variable, like some kind of "nucleus" of a figure that, however, will never appear before the eye because the figure is not the proper object of sight. We are going to call this organization of the immediately visible the "figural," insofar as the figural is precisely what is formless, the visibility of the formless.[93] This figural is what Berkeley calls specifically visible extension. It is the "organization" or "constellation" of the immediately visible: "or rather visible extension seems to be the coexistence of colours in ye mind."[94] It is a coexistence of colors that do not have a form and the instability of that coexistence. The figural is, therefore, the simultaneity of formless immediate objects of sight plus the rhythm of their motion. But because the formless figural is visible extension, but not spatial extension, because it does not have its own place, it is not tangible.

Spatial extension, by contrast, and everything that defines the spatial situation of an object is the proper object of touch, and not of the glance. "For Berkeley, it becomes relevant to ask is situation properly perceived by sight or by some other means, and his answer is that situation is the proper object of touch."[95] This means that magnitude, shape, extension, and distance are immediate objects of touch. Thus, distance reappears, but not where common sense would expect it—as an object of touch. Touch becomes the sense that has the experience of distance. "We have a faculty by means of which we can immediately apprehend distance and things placed at a distance, namely, touch."[96] However, by defining touch as the "sense of distance," Berkeley did not reintroduce the split between outside and inside. On the contrary, touch is the "sense of distance" because it can negate the difference between outside and inside, or because it can negate that "between." Touch is the "sense of distance" because it is not the sense of distance. This means distance as the proper object of touch can exist for touch only when it no longer exists as distance.

The body, which is the organ of the sense of touch, estimates distance: a body stands up, starts moving, travels a certain distance, and reaches another body. When moving toward the other body, it is unaware of the distance it will have to cover. It moves "blindly" because the glance does not comprehend distance. So the body does not know how long it will travel toward the other body. It will discover that only when it is finally able to touch the other body—when the distance between them disappears. The tactile object exists as an object as soon as it becomes possible to reach it, as soon as distance is annulled. Touch can determine distance only in the moment of its annulment, only in the moment of touching the tactile object. We feel touched by what we touch. Touch is, therefore, a sense that demands self-transcendence, a sense that is established by this self-abandoning and by entering the other. Touch does not exist without a mad intimacy with the other, without merging with the touched. It always gets into the other, thus disturbing boundaries between bodies. Touch is the disturbance of the configuration of constellations of sensations.

However, to say that to touch the other always means to be touched by the other is not to say that touch repeats the reflexive doubling inscribed in the gaze. Touch does not divide itself in order to touch itself. To be touched in the act of touching does not mean to retrieve oneself. On the contrary, it means letting oneself go outside oneself or letting oneself into the other that is touched: to destabilize, to disturb or to negate the "firm contours" of one's own identity (of one's own body). To put it simply, touch is a sense of pure openness. It keeps the "door" of the body constantly open. It is a sense of reception, of acceptance, of letting oneself be within the other. It is a sense of hospitality and of giving up (of transcending) one's own identity. Touch is therefore a paradoxical sense: it is the sense of distance insofar as it "reaches" the tactile object or arrives at it from a distance, and, at the same time, it is a sense without "the sense" of distance, insofar as touching is possible only through the annihilation of distance.

The touched body is a palpable extension. We repeat: palpable extension has nothing to do with visible extension: "there is no necessary connexion between those two distinct extensions."[97] The body that touches another body *feels* the other body as hard or soft, as warm or cold. The eye, however, will never experience this. Between the visible and the palpable there is no necessary connection. But it is not only that between the object of sight and the object of touch there is no necessary connection, it is that they do not resemble one another: "W^t I see is only variety of colors & light. w^t I feel is hard or soft, hot or cold, rough or smooth, &c. w^t resemblance have these thoughts with those?"[98] Since visible and palpable are not necessarily connected (or since they are heterogeneous), the only relation that can be established between them is "conventional," an artificial relation, the one that is, for instance, established between words and things. The visible could function as the sign of the palpable. Which is not to say that the palpable can be "translated" into the visible. The visible and the palpable are not two different languages connected by a common meaning. That would be possible if the visible could be established as "the proper name" of the palpable. However, this possibility is in fact a pure impossibility, for in Berkeley's conception of visual language, the visible functions only as its own proper name, which is to say as its "own" verb insofar as perception is what is perceived: a paradoxical entity that acts.

Every visible is, thus, a unique and irreducible verb that does not refer to anything—this is Berkeley's idea of a strict language without ambiguities, in which no verb mingles with another, in which every verb is nothing other than its own immediate presentation: "And this brings into prominence his notion of what a strict language should be. It would be a language in which to each verb of perception would be appropriated its own list of [grammatical] objects and adjectives; if any words appeared on more than one list they would be ambiguous words, with as many different senses as there are different lists on which they appear."[99] This means that the word appearing in two languages would have two irreducibly different senses, would double itself in two words. Therefore, the relation between visible and palpable is not the relation between two words of different languages that can be translated into one another. It is not the relation between words and things. Both the visible and the palpable are things, or word-verbs. It is always a matter of two different, nonconnected things or two different words. It is not possible to say that the visible *means* something that is palpable.

Thanks to this schism between the visible and the palpable, one cannot say that between them there is a relation analogous to that established by Vico in his elaboration of poetic or heroic language. The difference between Berkeley's and Vico's interpretation of the visible and the palpable remains irreducible, in spite of the similarities between their understanding of vision and the visible. For Vico, as well as for Berkeley, "vision, in itself, is a kaleidoscope of floating, random, disparate sensations," mutually unconnected. However, in contrast to Berkeley, Vico introduces "the third eye," the eye of ingenium, which will connect not only random visible sensations, but also the sensations of touch and hearing, thus establishing a unique, poetic language. Moreover, the "initial" sensation that will set in motion the procedure of connecting the sensations of different senses does not come from the visible: "the sensation that opened that poetic eye was that of hearing . . . it was the sound of thunder that forced the poets to raise their bodily eyes and see, for the first time, . . . the vast emptiness of the sky."[100] What the ear heard forced the eye to watch and to see the emptiness of sky, and this emptiness horrified the poet, causing the trembling of his body. And "the body that trembles" before the emptiness of the sky experiences that emptiness as the "alive body of God," which affects it and therefore touches it.

In this way, the connection of the visible, palpable, and audible emerges. They are different, but their differences are connected by the common meaning of a metaphor produced through imitation: the image that the eye sees is the visible metaphor of the audible image that the ear heard. The eye saw what the ear heard, and what the eye saw touched the body, thus forcing it to tremble. The trembling body now moves through the world, establishing by its movements "the mute language of gestures," a still-unarticulated language: "Those agitated gestures expressed in bodily writing the terror they felt: a writing that, as bodily *imitation of a metaphoric image*, which in turn signified that the image was . . . metaphoric."[101]

It is always a matter of the "translation" of metaphors: the image of empty sky is the metaphor of thunder that the ear heard, and the trembling body is a metaphor of the "alive body of God" that is the image of an empty sky. After the "initial event"—the sound of the thunder—the entire mechanism of imitation and expression is set in motion: the first audible image is "projected" into the sky, thus connecting hearing and sight. The effect of this connection was understood as a metaphor of the body of God. The life of this body is imitated by gestures of human bodies that are gestural metaphors of the visible that connect the visible and the palpable in a labor that, as the labor of metaphor, refers to a natural relation between what is different: thence comes "the vividness of heroic speech, which was the direct successor of the mute language of the divine age, which had conveyed ideas through gestures and objects naturally related to them," the vividness of a language in a natural relation with designated ideas and spoken in the age when heroes ruled.[102] The natural relation is the relation of similarity between differences, what enables them to imitate themselves and express themselves and one another through metaphor.[103] In other words, it enables them to articulate a unique, heroic language that, through this articulation, forsakes the confused chaos of ideas or sensations and so abandons Babylon.

For Berkeley, however, the connection of differences through a metaphor is impossible, because it presupposes the existence of the "eye of ingenium" that, of course, is not the Cartesian "epistemological eye," but is nevertheless conceived of as the "ontological power of corporeal soul"[104] and therefore as the eye that has the power to "see the invisible," to see abstractions. Yet the eye that sees abstractions is itself only an abstraction, something one must reject as a mere fiction. We are again facing the same question: If between the visible and the palpable a field of common meaning cannot be established, then what kind of relation can be formed between them? If language is understood as the force establishing the determined relation between what Berkeley calls "sign" and "signified," then in his "structure" of the relation between the visible and the palpable, the visible cannot be the sign of the palpable. However, if one can imagine a mad sign that has no determined (conventional) connection with its referent, a sign that always refers to some other referent, then, and only then, does the visible become the sign of the palpable.

This is the entire logic of the analogy with language introduced by Berkeley in order to articulate the relation between the visible and the palpable. When Berkeley says that between the visible and the palpable there might exist a relation similar to the one that exists in language between a sound that is beyond the meaning of the word transferred by this sound and that word, he does not refer to their interdependence. What is more, he does not refer to the possibility that such a connection can exist. He refers to the fact that it does not have to exist. He refers to the absolute accidentality of this connection, to the fact that what is connected by this connection can be absolutely different and therefore not connected. The visible and the palpable are absolutely different: "Ideas which constitute the tangible earth and man *are completely different* from those which constitute the visible earth and man." "We do not find there is any necessary connexion betwixt this or that tangible quality of any colour whatever. And we may sometimes perceive colours, where there is nothing to be felt."[105] The visible and the invisible are, therefore, completely different: they exist independently. The visible exists without the palpable, and vice versa. If one can speak at all about the relation between the visible and the palpable, it would, therefore, be formulated as the relation of free dependence, where two parts are in mutual connection, but in such a way that the first does not presuppose the other, nor the other the first. In other words, the visible and the palpable can constitute a constellation that is always provisional, changeable, and heterogeneous.

The fact that between the visible and the palpable a constellation can be created means that between them there can be established a relation that is always accidental and random—a relation of suggestion. The visible can be a suggestion to the touch and therefore a kind of "advice" or proposition, an indication, a tone, a wink, or a trace. However, this trace has the nature of any trace: it is not necessarily or causally connected to any other trace. From this trace nothing necessarily leads to "the culprit." Berkeley's idea of suggestion suggests, therefore, that suggestion does not have to offer a correct or a "good" trace, for it can in fact lead to a wrong trace: "But that one might be deceived by those suggestions of course, and that there is no necessary connexion between visible and tangible ideas suggested by them, we need go no further than the next looking-glass."[106]

The suggestion seduces and deceives: it leads in a wrong direction. A wrong direction, however, does not mean that there exists an untrue idea of vision or of touch. This kind of interpretation is impossible because an untrue idea does not exist at all. Every visible idea is true and becomes "erroneous" only if and when it is understood as suggestion, when it is connected with what is different from it, when it is understood as a heterogeneous constellation of visible and tangible. This is to say: the gaze sees the color yellow, but for touch, yellow is the suggestion or the trace, for example, of a lemon. The arm moves to take the lemon, and in the moment it touches it, it is hurt, for it does not touch a lemon at all, but an object that is burning. The arm touches fire. Each of these ideas is true: the visible idea of what is yellow is true, as well as the tangible idea of what is hot. The most heterogeneous and unpredictable constellations are possible. The visible idea of what is yellow can appear simultaneously with the palpable idea of a lemon or something on fire.

This is only another way of saying that the visible and the palpable are irreducibly different, that vision can never touch and that touch can never see: "Now bodies operating on our organs, by an immediate application, and the hurt or advantage arising therefrom, depend altogether on the tangible, and not at all on the visible, qualities of any kind."[107] Bodies are left to the randomness of encounters with other bodies, to the touching that remains "blind," always uncertain and unpredictable: the effect of touch can be pleasure or pain, joy or damage, the preservation or destruction of a constellation that constitutes the body. The eye, for its part, is left to pure visibility, to pictures that cannot be touched. That is why one can say that Berkeley has introduced the notion of suggestion in order to suggest the absence of relation between the ideas of different senses. In whatever way they may be connected, the visible and the palpable remain heterogeneous: "It must be acknowledged that we never see and feel one and the same object. That which is seen is one thing, and that which is felt is another. . . . The true consequence is that the objects of sight and touch are two distinct things."[108]

If the notion of the haptic is understood as a kind of priority given to touch over the gaze, if the haptic is understood to mean that touch is the criterion of truthfulness of the visible, then one cannot claim that Berkeley gave a priority to the haptic over the optic. The heterogeneity of the visible and the tangible means that the visible is its own truth, that it is not subordinated to touch. However, one can say that in Berkeley, a priority of the haptic is established on the condition that the haptic be understood differently, on the condition that it be understood as the heterogeneity of the visible and the palpable. In this heterogeneous haptic space, the touch of the hand is neither a negation nor a confirmation of the visible. Here, the hand "works" for itself and the gaze sees for itself. "To characterize the connection of eye and hand, it is certainly not enough to say that the eye is infinitely richer, and passes through dynamic tensions, logical reversals and vicariances. . . . We will speak of the *haptic* each time there is no longer strict subordination in one direction or the other,"[109] which means each time the eye is not subordinated to the hand or each time the hand becomes "disobedient," each time the heterogeneity between the eye and the hand is maintained.

In other words, the heterogeneity between sight and touch is possible only within a haptic space that is heterogeneous within itself. This does not mean only that a hand touches a body in haptic space. It also means that an eye can see a picture only in haptic space. Sight, as understood by Berkeley, does not know of optical space. Berkeleian vision is haptic: the glance is possible only in a haptic space where distance is not the proper object of sight. The haptic eye can exist only where distance is imperceptible: "Where there is close vision, space is not visual, or rather the eye itself has a haptic, nonoptical function: no line separates earth from sky, which are of the same substance; there is neither horizon nor background nor perspective nor limit nor outline or form, nor center; there is no intermediary distance, or all distance is intermediary."[110] When the optical space that relies on distances and horizons disappears, there appears the haptic eye, whose objects of sight are only colors. In haptic visible space, everything "goes towards color, in color." In this space, there are only modulations of colors and the rhythm of their change or repetition.

That is why the haptic eye announces a catastrophe of the optical. For the haptic eye, "the foreground form falls. . . . The foreground form is no longer essence."[111] When the eye becomes haptic, then depth becomes tactile. Then a hand touches a body unknown to the eye in a space through which it moves "blindly." The blind body that is the sense of touch thus becomes "the function" of the haptic. It moves through a haptic space "guided" only by the space that is without fore*seeable* marks. In haptic space, every encounter between two bodies is therefore a crash. "Contrary to what is sometimes said, one never sees from a distance in a space of this kind, nor does one see it from a distance; one is never 'in front of,' any more than one is 'in'. . . . Orientations are not constant but change according to temporary vegetation, occupations, and precipitation. There is no visual model for points of reference."[112] The hand penetrates a space that cannot be seen and comes across another body that was not announced to it by any sign or suggestion. In this way, the hand is left to encounters and becomings that remain invisible to the eye, and vice versa, the eye knows of becomings that the hand cannot touch. The hand travels alone, as it were, following its own trajectory, and the eye travels for itself. Each of them moves toward different and heterogeneous becomings.

The circle is once more closed—and again we encounter schizophrenia. Schizophrenia implies the heterogeneity of immediate objects of different senses. As a schizophrenic says: "'A bird is twittering in the garden. I can hear the bird and I know that it is twittering, but that it is a bird and that it is twittering, two things seem so remote from each other. . . . There is a gulf between them, as if the bird and the twittering had nothing to do with each other.'"[113]

The world of a schizophrenic splits, its unity fissured by the abysses between sound and shape, between color and distance, between depth and form—they do not merge into a whole object. The world is disjointed because the body "has ceased to draw together all objects in its one grip."[114] Every sense goes its own way, to its own objects. This indeed is schizophrenia, not madness, not illness, and certainly not a nervous breakdown, but the breakthrough that establishes and preserves the schizophrenically "unified," heterogeneous object. By crossing this limit, the iconographic subject simultaneously enters different objects. It is at every moment multiple—it is the multitude of objects, of becomings, of comings and leavings. It is color, it is sound, it is pain and pleasure. "Useless to ask (as *we* should) what was really there, or what actually happened; it could only be said that a certain sound, a pattern of colours was seen, and a shock was felt."[115]

A world of nightmare appears in which only surprises and astonishments are continuous: a world of horror, a world of schizophrenia. "These men . . . or do they not yet exist?—are like Zarathustra. They know incredible sufferings, vertigos, and sicknesses. They have their specters. They must reinvent each gesture."[116] They have to do that because they are in a haptic space that is without marks of direction and because they never know where they are going and how to proceed. They have their specters, but these specters are not ghosts that produce the world of identities. These are specters of constant horror and astonishment, of unbearable fear that is the effect of being willing to accept everything at any price and to become everything. This acceptance is nothing other than a constant crossing of the border. It is becoming as "leaving from." The schizophrenic, iconographic subject has to do that time and again: he has to cross the line. "He has crossed over the limit, the schiz. . . . The schizo knows how to leave: he has made departure something as simple as being born or dying."[117] The iconographic subject knows what is most difficult—he knows how to leave.

In this haptic world, nothing can be connected. Here everything is heterogeneous, everything presents only itself as its own name-verb. There is no object that could gather differences, and there is no name that could name differences. That every object bears or is its own name that is a verb means no one in this world can be understood, for there is nothing but a proper name, "the absolute idiom."[118] In God's book there is no word that could connect other words. God's visual language, therefore, "annihilates the gift of language, or at least makes a babel of it." God's visual language is confusion. Whenever we make an attempt to connect two ideas or two names, whenever we understand a picture as a suggestion of some other picture, a *malentendu* appears, which in the Babylonian multitude of languages means "to mishear as to misunderstand."[119] Every connection is such a misunderstanding. Everything remains a blunder. Once more, and for the last time, the circle is closed. The name of this confusion, the name of divine language, is some form of Babel or Babylon, the name of God as the name of the Father: "God proclaimed his name loudly, the name which he himself has chosen and which is thus his." However, Babel means not only the name of God, but also confusion: "He imposes confusion on them at the same time as he imposes his proper name, the name he has chosen which means confusion, which seems confusedly to mean confusion."[120] By giving his name, by giving all names, the father would find himself at the origin of language. God's visual language is a state of babble caused by constant interruptions, the confusion of the absence of causal connections. The origin of languages—Babel—names the name of this origin: "And the name of God the Father would be the name of that origin of language. . . . "

APPENDIX

Samuel Beckett, *"Film"*

Throughout first two parts all perception is E's. E is the camera. But in third part there is O's perception of room and contents and at the same time E's continued perception of O. This poses a problem of images which I cannot solve without technical help.

The film is divided into three parts. 1. The street (about eight minutes). 2. The stairs (about five minutes). 3. The room (about seventeen minutes).

The film is entirely silent except for the 'sssh!' in part one.

Climate of film comic and unreal. O should invite laughter throughout by his way of moving. Unreality of street scene . . .

GENERAL

Esse est percipi

All extraneous perception suppressed, animal, human, divine, self-perception maintains in being.

Search of non-being in flight from extraneous perception breaking down in inescapability of self-perception.

No truth value attaches to above, regarded as of merely structural and dramatic convenience.

In order to be figured in this situation the protagonist is sundered into object (O) and eye (E), the former in flight, the latter in pursuit.

It will not be clear until end of film that pursuing perceiver is not extraneous, but self.

Until end of film O is perceived by E from behind and at an angle not exceeding 45°. Convention: O enters *percipi* = experiences anguish of perceivedness, only when this angle is exceeded.

E is therefore at pains, throughout pursuit, to keep within this 'angle of immunity' and only exceeds it (1) inadvertently at beginning of part one when he first sights O (2) inadvertently at beginning of part two when he follows O into vestibule and (3) deliberately at end of part three when O is cornered. In first two cases he hastily reduces angle.

OUTLINE

The street

Dead straight. No sidestreets or intersections. Period: about 1929. Early summer morning. Small factory district. Moderate animation of workers going unhurriedly to work. All going in same direction and all in couples. No automobiles. Two bicycles ridden by men with girl passenger (on crossbar). One cab, cantering nag, driver standing brandishing whip. All persons in opening scene to be shown in some way perceiving—one another, an object, a shop window, a poster, etc., i.e. all contentedly in *percipere* and *percipi*. First view of above is by E motionless and searching with his eyes for O. He may be supposed at street edge of wide (4 yards) sidewalk. O finally comes into view hastening blindly along sidewalk, hugging the wall on his left, in opposite direction to all the others. Long dark overcoat (whereas all others in light summer dress) with collar up, hat pulled down over eyes, briefcase in left hand, right hand shielding exposed side of face. He storms along in comic foundered precipitancy. E's searching eye, turning left from street to sidewalk, pick him up at an angle exceeding that of immunity (O's unperceivedness according to convention). O, entering perceivedness, reacts (after just sufficient onward movement for his gait to be established) by halting and cringing aside towards wall. E immediately draws back to close the angle and O released from perceivedness, hurries on. E lets him get about 10 yards ahead and then starts after him. Street elements from now on incidental (except for episode of couple) in the sense that only registered in so far as they happen to enter field of pursuing eye fixed on O.

Episode of couple. In his blind haste O jostles an elderly couple of shabby genteel aspect, standing on sidewalk, peering together at a newspaper. They should be discovered by E a few yards before collision. The woman is holding a pet monkey under her left arm. E follows O an instant as he hastens blindly on, then registers couple recovering from shock, comes up with them, passes them slightly and halts to observe them. Having recovered they turn and look after O, the woman raising a lorgnon to her eyes, the man taking off his pince-nez fastened to his coat by a ribbon. They then look at each other, she lowering her lorgnon, he resuming his pince-nez. He opens his mouth to vituperate. She checks him with a gesture and soft 'sssh!' He turns again, taking off his pince-nez, to look after O. She feels the gaze of E upon them and turns, raising her lorgnon, to look at him. She nudges her companion who turns back towards her, resuming his pince-nez, follows direction of her gaze and, taking off his pince-nez, looks at E. As they both stare at E the expression gradually comes over their faces which will be that of the flower-woman in the stairs scene and that of O at the end of film, an expression only to be described as corresponding to an agony of perceivedness. Indifference of monkey, looking up into face of its mistress. They close their eyes, she lowering her lorgnon, and hasten away in direction of all the others, i.e. that opposed to O and E.

E turns back towards O by now far ahead and out of sight. Immediate acceleration of E in pursuit (blurred transit of encountered elements). O comes into view, grows rapidly larger until E settles down behind him at same angle and remove as before. O disappears suddenly through open housedoor on his left. Immediate acceleration of E who comes up with O in vestibule at foot of stairs.

Stairs

Vestibule about 4 yards square with stairs at inner righthand angle. Relation of streetdoor to stairs such that E's first perception of O (E near door, O motionless at foot of stairs, right hand on banister, body shaken by panting) is from an angle a little exceeding that of immunity. O, entering perceivedness (according to convention), transfers right hand from banister to exposed side of face and cringes aside towards wall on his left. E immediately draws back to close the angle and O, released, resumes his pose at foot of stairs, hand on banister. O mounts a few steps (E remaining near door), raises head, listens, redescends hastily backwards and crouches down in angle of stairs and wall on his right, invisible to one descending. E registers him there, then transfers to stairs. A frail old woman appears on bottom landing. She carries a tray of flowers slung from her neck by a strap. She descends slowly, with fumbling feet, one hand steadying the tray, the other holding the banister. Absorbed by difficulty of descent she does not become aware of E until she is quite down and making for the door. She halts and looks full at E. Gradually same expression as that of couple in street. She closes her eyes, then sinks to the ground and lies with face in scattered flowers. E lingers on this a moment, then transfers to where O last registered. He is no longer there, but hastening up the stairs. E transfers to stairs and picks up O as he reaches first landing. Bound forwards and up of E who overtakes O on second flight and is literally at his heels when he reaches second landing and opens with key door of room. They enter room together, E turning with O as he turns to lock the door behind him.

The room

Here we assume problem of dual perception solved and enter O's perception. E must so manoeuvre throughout what follows, until investment proper, that O is always seen from behind, at most convenient remove, and from an angle never exceeding that of immunity, i.e. preserved from perceivedness.

Small barely furnished room. Side by side on floor a large cat and small dog. Unreal quality. Motionless till ejected. Cat bigger than dog. On a table against wall a parrot in a cage and a goldfish in a bowl. This room sequence falls into three parts.

1. Preparation of room (occlusion of window and mirror, ejection of dog and cat, destruction of God's image, occlusion of parrot and goldfish).
2. Period in rocking-chair. Inspection and destruction of photographs.
3. Final investment of O by E and dénouement.

1. O stands near door with case in hand and takes in room. Succession of images: dog and cat, side by side, staring at him; mirror; window; couch with rug; dog and cat staring at him; parrot and goldfish, parrot staring at him; rocking-chair; dog and cat staring at him. He sets down case, approaches window from side and draws curtain. He turns towards dog and cat, still staring at him, then goes to couch and takes up rug. He turns towards dog and cat, still staring at him. Holding rug before him he approaches mirror from side and covers it with rug. He turns towards parrot and goldfish, parrot still staring at him. He goes to rocking-chair, inspects it from front. Insistent image of curiously carved headrest. He turns towards dog and cat still staring at him. He puts them out of room. He takes up case and is moving towards chair when rug falls from mirror. He drops briefcase, hastens to wall between couch and mirror, follows walls past window approaches mirror from side, picks up rug and, holding it before him, covers mirror with it again. He returns to briefcase, picks it up, goes to chair, sits down and is opening case when disturbed by print, pinned to wall before him, of the face of god the Father, the eyes staring at him severely. He sets down case on floor to his left, gets up and inspects print. Insistent image of wall,

paper hanging off in strips. He tears print from wall, tears it in fours, throws down the pieces and grinds them underfoot. He turns back to chair, image again of its curious headrest, sits down, image again of tattered wall-paper, takes case on his knees, takes out a folder, sets down case on floor to his left and is opening folder when disturbed by parrot's eye. He lays folder on case, gets up, takes off overcoat, goes to parrot, close up of parrot's eye, covers cage with coat, goes back to chair, image again of headrest, sits down, image again of tattered wall-paper, takes up folder and is opening it when disturbed by fish's eye. He lays folder on case, gets up, goes to fish, close-up of fish's eye, extends coat to cover bowl as well as cage, goes back to chair, image again of headrest, sits down, image again of wall, takes up folder, takes off hat and lays it on case to his left. Scant hair or bald to facilitate identification of narrow black elastic encircling head.

When O sits up and back his head is framed in headrest which is a narrower extension of backrest. Throughout scene of inspection and destruction of photographs E may be supposed immediately behind chair looking down over O's left shoulder.

2. O opens folder, takes from it a packet of photographs, lays folder on case and begins to inspect photographs. He inspects them in order 1 to 7. When he has finished with 1 he lays it on his knees, inspects 2, lays it on top of 1, and so on, so that when he has finished inspecting them all 1 will be at the bottom of the pile and 7—or rather 6, for he does not lay down 7—at the top. He gives about six seconds each to 1–4, about twice as long to 5 and 6 (trembling hands). Looking at 6 he touches with forefinger little girl's face. After six seconds of 7 he tears it in four and drops pieces on floor on his left. He takes up 6 from top of pile on his knees, looks at it again for about three seconds, tears it in four and drops pieces on floor to his left. So on for the others, looking at each again for about three seconds before tearing it up. 1 must be on tougher mount for he has difficulty in tearing it across. Straining hands. He finally succeeds, drops pieces on floor and sits, rocking slightly, hands holding armrests.

3. Investment proper. Perception from now on, if dual perception feasible, E's alone, except perception of E by O at end. E moves a little back (image of headrest from back), then starts circling to his left, approaches maximum angle and halts. From this open angle, beyond which he will enter *percipi*, O can be seen beginning to doze off. His visible hand relaxes on armrest, his head nods and falls forward, the rock approaches stillness. E advances, opening angle beyond limit of immunity, his gaze pierces the light sleep and O starts awake. The start revives the rock, immediately arrested by foot to floor. Tension of hand on armrest. Turning his head to right O cringes away from perceivedness. E draws back to reduce the angle and after a moment, reassured, O turns back front and

resumes his pose. The rock resumes, dies down slowly as O dozes off again. E now begins a much wider encirclement. Images of curtained window, walls and shrouded mirror to indicate his path and that he is not yet looking at O. Then brief image of O seen by E from well beyond the angle of immunity, i.e. from near the table with shrouded bowl and cage. O is now seen to be fast asleep, his head sunk on his chest and his hands, fallen from the armrests, limply dangling. E resumes his cautious approach. Images of shrouded bowl and cage and tattered wall adjoining, with same indication as before. Halt and brief image, not far short of full-face, of O still fast asleep. E advances last few yards along tattered wall and halts directly in front of O. Long image of O, full-face, against ground of headrest, sleeping. E's gaze pierces the sleep, O starts awake, stares up at E. Patch over O's left eye now seen for the first time. Rock revived by start, stilled at once by foot to ground. Hand clutches armrest. O half starts from chair, then stiffens, staring up at E. Gradually that look. Cut to E, of whom this very first image (face only, against ground of tattered wall). It is O's face (with patch) but with very different expression, impossible to describe, neither severity nor benignity, but rather acute *intentness*. A big nail is visible near left temple (patch side). Long image of the unblinking gaze. Cut back to O, still half risen, staring up, with that look. O closes his eyes and falls back in chair, starting off rock. He covers his face with his hands. Image of O rocking, his head in his hands but not yet bowed. Cut back to E. As before. Cut back to O. He sits, bowed forward, his head in his hands, gently rocking. Hold it as the rocking dies down.

END

Notes

Chapter 1. The Passive Synthesis of Contemplation

1. Bruno, *Cause, Principle and Unity*, p. 36.
2. Ibid., p. 37.
3. Ibid., p. 38.
4. Bruno, "Shadows" (Op. lat. II (i), p. 9). Quoted in Yates, *The Art of Memory*, p. 217.
5. For analysis of the oppositional metaphor see Hocke, *Manierismus*, pp. 61–63: "Metaphor, the transportation of one thing into the other, assumes in mannerism the importance of the means of communication that is in perfect accordance with a world without structure; what is more, it assumes the meaning of a miracle. The infinite play of transformation allowed by metaphor . . . as developed in mannerism attains the meaning of the mirror of the world. . . . The art of metaphor is the root of all arts. Oppositional metaphor, the one that connect oppositions, is the best product of wisdom for it is something—and this is important—that rhetoricians are ignorant of."
6. Bruno, *Cause, Principle and Unity*, p. 18.
7. Ibid.
8. Ibid., p. 41.
9. Yates, *The Art of Memory*, p. 212.
10. Bruno, *The Heroic Frenzies*, second part, second dialogue.
11. Bruno, *Cause, Principle and Unity*, p. 65.
12. Ibid.
13. Schelling, *Bruno*, p. 160.
14. Ibid., p. 162.
15. Yates, *The Art of Memory*, p. 216.
16. See Godwin, *Athanasius Kircher*, p. 72.
17. Bruno, "Essays on Magic," p. 137.
18. Ibid.
19. Ibid., p. 138.
20. Ibid., p. 137.
21. Bruno, *The Heroic Frenzies*, second part, second dialogue.

22. Ibid.
23. Ibid.
24. Ibid., first part, first dialogue.
25. Ibid., "Argument of the Nolan."
26. Ibid., second part, fourth dialogue.
27. Ibid., first part, fourth dialogue.
28. Ibid., second part, fourth dialogue.
29. Hocke, *Manierismus in der Literatur*, p. 131: "Writers of mannerstic treatises celebrate the elegance of surprise that occurs in wrong conclusions, the shocking effect of an unexpected turn. However, they celebrate most . . . peripety, 'the surprising turn.'"
30. Bruno, *The Heroic Frenzies*, second part, fourth dialogue.
31. Ibid.
32. Ibid.
33. Ibid.
34. Ibid.
35. Ibid.
36. Ibid.
37. Ibid.
38. Ibid.
39. Ibid.
40. Foucault, *The Order of Things*, p. 24.
41. Bruno, *The Heroic Frenzies*, second part, second dialogue.
42. Ibid.
43. Ibid.

Chapter 2. The Active Synthesis of Reflection

1. Descartes, "The World, or Treatise on Light," p. 89.
2. Ibid.
3. Ibid.
4. Ibid., pp. 84–85.
5. Descartes, "Optics," p. 153. This transfer is instantaneous. Light is transferred in a moment from the object to the eye. See Sabra, *Theories of Light from Descartes to Newton*, pp. 46–48.
6. Lacan, *The Four Fundamental Concepts of Psychoanalysis*, p. 75.
7. Descartes, "Optics," p. 154.
8. Descartes, "Le Monde, Traité de l'Homme," p. 186.
9. Ibid.
10. Descartes, "Optics," p. 165.
11. Merleau-Ponty, "Eye and Mind," p. 131.
12. Descartes, "The World, or Treatise on Light," p. 81.
13. Ibid.

14. Ibid.
15. Foucault, *Death and the Labyrinth*, p. 112.
16. Descartes, "Optics," pp. 168–69.
17. Ibid., pp. 165–66.
18. Ibid., p. 165.
19. Foucault, *The Order of Things*, p. 64.
20. Descartes, "Treatise on Man," p. 101.
21. Merleau-Ponty, "Eye and Mind," p. 389.
22. Descartes, "Optics," p. 165.
23. Deleuze/Guattari, *A Thousand Plateaus*, p. 77.
24. Ibid., p. 115.
25. Deleuze, *Cinema 1: The Movement-Image*, p. 88.
26. Descartes, "Treatise on Man," p. 106.
27. Descartes, "The Passions of the Soul," p. 353.
28. Ibid.
29. Lyotard, "Scapeland," p. 186.
30. Descartes, "Optics," pp. 154–55.
31. Ibid., p. 155.
32. Lacan, *The Four Fundamental Concepts of Psychoanalysis*, p. 86.
33. Ibid., p. 93.
34. Descartes, "Optics," p. 153.
35. Merleau-Ponty, "Eye and Mind," p. 131.
36. Descartes, "Optics," p. 170.
37. Merleau-Ponty, *The Visible and the Invisible*, p. 7.
38. Lacan, *The Four Fundamental Concepts of Psychoanalysis*, p. 86.
39. Deleuze, "On the Movement-Image," p. 54.
40. Merleau-Ponty, *Phenomenology of Perception*, p. 216.
41. Ibid., p. 217.
42. Descartes, "Optics," p. 172.
43. It now becomes clear why Descartes saw the entire ground of his philosophy and, first of all, his metaphysics, in his optics. It becomes clear why he thought that his entire "philosophy would be shaken to the ground" (*Correspondance*, p. 308) if the theory that light is momentarily projected from object to the eye and further to the brain were to be experimentally refuted. For the importance of Descartes's conception of projection for the problem of the "new subjectivity" introduced in his *Meditations*, see Blumenberg, "Light as a Metaphor for Truth," p. 39.
44. Foucault, *The Order of Things*, pp. 15–16.
45. Descartes, "Early Writings," p. 2.
46. Damisch, *The Origin of Perspective*, p. 46. See also Goux, "Descartes et la perspective."
47. Damisch, *The Origin of Perspective*, p. 51.
48. Descartes, "Principles of Philosophy," p. 232, par. 21.

49. Swift, *Gulliver's Travels*, p. 37.
50. Michel Serres, "Le Système de Leibniz et ses modèles mathémathiques" (Paris, 1968), p. 693, quoted in: Damisch, *The Origin of Perspective*, p. 49.
51. Merleau-Ponty, *The Visible and the Invisible*, p. 210.
52. Ibid., p. 248.
53. Copjec, *Read My Desire*, p. 124.
54. Pascal, *Pensées*, par. 445.
55. Pascal, "De l'Esprit géometrique," pp. 590–91.
56. Lacan, *The Seminar of Jacques Lacan: Book II*, p. 6.
57. Ibid., p. 50.
58. Descartes, "Rules for the Direction of the Mind," p. 41, Rule XII.
59. Bergson, *Creative Evolution*, p. 328.
60. Descartes, "Le Monde, Traité de l'Homme," p. 184.
61. Bergson, *Creative Evolution*, p. 328.
62. Descartes, "Le Monde, Traité de l'Homme," p. 158.
63. Bergson, *Creative Evolution*, p. 333.
64. Descartes. "The Search After Truth," p. 312.
65. Calderon de la Barca, *Life is a Dream*, p. 345.
66. Descartes, "Le Monde, Traité de l'Homme," p. 198.
67. Ibid.

Chapter 3 (God)

1. Berkeley, "An Essay Towards a New Theory of Vision," p. 9, par. 12.
2. Ibid.
3. Berkeley, "The Theory of Vision Vindicated and Explained," p. 296, par. 50.
4. Ibid.
5. "Both Descartes and Locke, along with Malebranche, Leibniz, and a host of others, believe that the sensory world we experience is wholly different from the material world that gives rise to it, our perceptions do not mirror nature at all." The world, therefore, remains invisible. See Wilson, "Discourses of Vision in Seventeenth Century Metaphysics," p. 122. See also Rorty, *Philosophy and the Mirror of Nature*, p. 68: "Descartes' only improvement on the Homeric notion of an Invisible and Intangible Man was to strip the intruder of humanoid form."
6. Berkeley, "The Theory of Vision Vindicated and Explained," p. 296, par. 50.
7. For the analysis of copy as the "good image," see Deleuze, "The Simulacrum and Ancient Philosophy," p. 257: "For if copies or icons are good images and are well-founded, it is because they are endowed with resemblance. But resemblance should not be understood as an external relation. It goes less from one thing to another than from one thing to an Idea."
8. Berkeley, "The Eruption of Mount Vesuvius," p. 247.
9. Berkeley, "The Theory of Vision Vindicated and Explained," p. 297, par. 51.

10. Berkeley, "An Essay Towards a New Theory of Vision," p. 50, par. 117.
11. Berkeley, *The Principles of Human Knowledge*, p. 67, par. 5.
12. Deleuze, "Simulacrum and Ancient Philosophy," p. 258. For an analysis of the simulacrum as the object that escapes representation and causality, see Lyotard, "Acinema," p. 171: "A simulacrum, understood in the sense Klossowski gives it, should not be conceived primarily as belonging to the category of representation."
13. Copjec, *Read My Desire*, p. 164.
14. Berkeley, *The Principles of Human Knowledge*, p. 97, par. 65.
15. Berkeley, *Philosophical Commentaries*, p. 54, par. 433.
16. Ibid., p. 102, par. 856.
17. Ibid., p. 75, par. 606.
18. Deleuze, "The Simulacrum and Ancient Philosophy," p. 258.
19. Berkeley, "The Theory of Vision Vindicated and Explained," p. 292, par. 38.
20. Foucault, *This is Not a Pipe*, p. 24.
21. Barthes, "On Photography," p. 353.
22. Foucault, *This is Not a Pipe*, p. 38.
23. Ibid., p. 195.
24. Ibid., p. 33.
25. Berkeley, *Philosophical Commentaries*, p. 94, par. 780.
26. Deleuze, *Difference and Repetition*, p. 126.
27. Deleuze, *The Logic of Sense*, p. 21. See also p. 145: "By itself, representation is given up to an extrinsic relation of resemblance or similitude only."
28. Berkeley, *The Principles of Human Knowledge*, p. 80, par. 65: "To all which my answer is, first, that the connexion of ideas does not imply the relation of cause and effect, but only of a mark or sign with the thing signified."
29. Berkeley, "Alciphron, or the Minute Philosopher," p. 149.
30. "Berkeley believes that neither persons nor things have identity, if identity means durational continuity." G. A. Johnston, editor's notes in Berkeley, *Commonplace Book*, p. 125.
31. Berkeley, "Siris," p. 99, par. 198. From the universal culinary fire there emerge bodies as provisional constellations of sensations, as culinary fires of "smaller" intensity: "Every ignited body is, in the foregoing sense, culinary fire."
32. Artaud, "Van Gogh," p. 499.
33. Deleuze, *The Logic of Sense*, p. 91.
34. Ibid., p. 91. For Deleuze's analysis of Lewis Carroll's "dilemma," see also pp. 23–24.
35. Berkeley. *The Principles of Human Knowledge*, p. 60. par. 38.
36. Deleuze, *The Logic of Sense*, p. 24. See also A. A. Luce, *Berkeley and Malebranche*, p. 99: "Berkeley's idea is presentative. It is not, like Malebranche's idea, 'different from' the thing.... The idea is just what it pretends to be, a particular. This idea is not that idea."
37. Deleuze/Guattari. *A Thousand Plateaus*, p. 313.

38. Ibid. p. 313.
39. Lowry, *Under The Volcano*, p. 136.
40. Serres, *Genesis*, p. 18.
41. Berkeley, "The Theory of Vision Vindicated and Explained," p. 287, par. 21.
42. Ibid., p. 286, par. 20: "As to the outward cause of these ideas, whether it be one and the same, or various and manifold, whether it be thinking or unthinking, spirit or body, or whatever else we conceive or determine about it, the visible appearances do not alter their nature, our ideas are still the same."
43. Ibid., p. 289, par. 30.
44. Ibid., p. 284, par. 11.
45. Ibid., p. 293, par. 40.
46. Lacan, *The Four Fundamental Concepts of Psychoanalysis*, p. 22.
47. Ibid.
48. Ibid., p. 45.
49. Copjec, *Read My Desire*, p. 131.
50. Berkeley. "Alciphron, or the Minute Philosopher," p. 149.
51. Copjec. *Read My Desire*, p. 34.
52. Berkeley. *The Principles of Human Knowledge*, p. 143, par. 151.
53. Bergson, *Creative Evolution*, p. 327: "In vain . . . shall we seek beneath the change the thing which changes: it is always provisionally." We cannot accept the thesis according to which the words of God's visual language are adjectives insofar as adjectives, even if they are "free" from nouns, still open a possibility for the existence of nouns: "The world of Berkeley has been accused of being a world of loose transferable adjectives, with no substantive nouns." See Ritchie, *George Berkeley*, p. 87. The idea that Berkeley's ideas have "*gappy existence*" is maybe closer to our thesis that they are verbs. See J. Dancy, *Berkeley: An Introduction*, p. 65.
54. Berkeley, *Philosophical Commentaries*, p. 78, par. 642: "The chief thing I do or pretend to do is onely to remove the mist or veil of Words."
55. Copjec, *Read My Desire*, pp. 188–89. For the analysis of the paradox of the locked room see pp. 190–96.
56. Berkeley, "Siris," p. 157, par. 349.
57. Deleuze, *The Logic of Sense*, p. 3.
58. Berkeley, *Philosophical Commentaries*, p. 34, par. 280.
59. Berkeley, "Siris," p. 81, par. 148.
60. Ibid, p. 106, par. 218. There are great and small evacuations of elements. Great evacuations are like grave diseases. They weaken the nature and can cause a disaster, the dissemination of all constellations into the amorphous culinary fire: "great evacuations weaken nature as well as the disease."
61. Berkeley, *Philosophical Commentaries*, p. 9, par. 12: "Revolutions immediately measure train of ideas, mediately duration." This is to say that revolutions can speed up or slow down the succession of ideas. They can speed up or slow down time so that by a revolutionary turn the consequence can precede its "own" cause.

62. Berkeley, "Siris," p. 86, par. 164. We will discuss the problem of desire in the next chapter.
63. Ibid., p. 86, par. 165.
64. Ibid., p. 100, par. 201.
65. Ibid., p. 108, par. 224.
66. Ibid., p. 88, par. 170.
67. Serres, *Genesis*, p. 109: "Chaos appears there, spontaneously, in the order, order appears there in the midst of disorder."
68. Berkeley, "Siris," p. 80, par. 145.
69. Ibid., p. 129, par. 273.
70. Ibid., p. 82, par. 152.
71. Ibid., p. 129, par. 273.
72. Ibid., p. 131, par. 279.
73. Ibid., p. 84, par. 158.
74. Ibid., p. 115, par. 239.
75. Berkeley, *Philosophical Commentaries*, p. 65, par. 522.
76. Ibid., p. 39, par. 318.
77. Berkeley, *The Principles of Human Knowledge*, p. 52, par. 12. For an analysis of the 'particular, general, abstract', see Burnyeat, "Idealism and Greek Philosophy," p. 25.
78. Berkeley, *The Principles of Human Knowledge*, p. 55, par. 15.
79. Berkeley, *Philosophical Commentaries*, p. 51, par. 406.
80. Ibid., p. 53, par. 422. See also p. 78, par. 638: "I affirm 'tis manifestly absurd— no excuse in yᵉ world can be given why a man should use a word without an idea." For a detailed analysis of this thesis and therefore of the problem it caused Berkeley in his attempt to maintain the possibility of founding the scientific laws, see Pitcher, *Berkeley*, p. 181.
81. Badiou, "The Fold," p. 58: "Leibniz-Deleuze says: the Multiple exists by concept, or: Multiple exists *in the One*. . . . This will establish . . . equivocity between 'to be an element of,' or 'belong to,' ontological categories, and 'to possess a property,' 'have a certain predicate,' categories of knowledge."
82. Berkeley, *Philosophical Commentaries*, p. 78, par. 642: "The chief thing I do or pretend to do is onely to remove the mist or veil of Words. This has occasion'd Ignorance & confusion. This has ruin'd the Scholemen and Mathematicians, Lawyers and Divines."
83. Ibid., p. 84, par. 693.
84. Deleuze, *Difference and Repetition*, pp. 34–35: "Here we rediscover the necessarily quadripartite character of representation."
85. Berkeley, *Philosophical Commentaries*, p. 90, par. 739.
86. Ibid., p. 12, par. 51.
87. Ibid., p. 86, par. 703.
88. Ibid., p. 89, par. 731.
89. Ibid., p. 89, par. 731a.

90. Ibid., p. 103, par. 873.
91. Ibid., pp. 84–85, par. 693.
92. Deleuze, *Cinema 2*, p. 130.
93. Berkeley, *Philosophical Commentaries*, p. 85, par. 696.
94. Berkeley, "Siris," p. 136, par. 292.
95. Deleuze, *Cinema 2*, p. 133.
96. Berkeley, *Philosophical Commentaries*, p. 89, par. 733.
97. Ibid., p. 90, par. 737: "To view the deformity of Errour we need onely undress it."
98. Deleuze, *Cinema 2*, p. 126.
99. Ibid., pp. 126–27.
100. Berkeley, *Philosophical Commentaries*, p. 71, par. 568.
101. Lacan, *The Seminar of Jacques Lacan, Book I*, p. 76.
102. Ibid., p. 77.
103. Deleuze, *Cinema 2*, p. 127.
104. Berkeley, "Siris," p. 114, par. 238.
105. Ibid., p. 114, par. 238.
106. Berkeley, *The Principles of Human Knowledge*, p. 145, par. 155.
107. Lacan, *The Seminar of Jacques Lacan, Book I*, p. 76.
108. Berkeley, "An Essay Towards a New Theory of Vision," p. 7, par. 2.
109. Berkeley, "De motu," p. 269, par. 52. Of course, Berkeley here has in mind first of all Newton and his distinction between relative and absolute space. Descartes also distinguished place from space, but his distinction differs somewhat from Berkeley's: "The difference between the terms 'place' and 'space' is that the former designates more explicitly the position, as opposed to the size or shape." On the other hand, "terms 'place' and 'space' . . . do not signify anything different from the body which is said to be in a place, nothing different from the body for which we say that it occupies the place." Descartes, "The Principles of Philosophy," pp. 229 and 228.
110. Berkeley, "De motu," p. 269, par. 53.
111. Ibid., p. 272, par. 59.
112. Leibniz, "An Example of Demonstrations About the Nature of Corporeal Things," p. 142.
113. Leibniz, "Letter to Jacob Thomasius," p. 97.
114. Derrida, *Memoirs of the Blind*, p. 44.
115. Leibniz, "Letter to Jacob Thomasius," p. 97.
116. Leibniz, "The Monadology," p. 545, par. 60.
117. Uspensky, *The Semiotics of the Russian Icon*, p.31.
118. Leibniz, "The Monadology," p. 545, par. 60.
119. Uspensky, *The Semiotics of the Russian Icon*, p. 33.
120. Ibid., pp. 33–34.
121. Florensky, *Iconostasis*, p. 65.
122. Ibid. For the "icon" as a symptom that cannot be controlled by a "central" per-

spective, but always returns to destabilize this perspective anew, see also W. J. T. Mitchell, *Iconology*, pp. 6 and 158.

123. Serres, "Panoptic Theory," p. 39.
124. Leibniz, "Tentamen Anagogicum," p. 479.
125. Ibid., p. 479.
126. Ibid., p. 480.
127. Serres, *Genesis*, p. 19.
128. Berkeley, "An Essay Towards a New Theory of Vision," p. 7, par. 2.
129. Serres, *Genesis*, p. 19.
130. Krauss, *The Optical Unconscious*, pp. 16–19.
131. Berkeley, *The Principles of Human Knowledge*, p. 145, par. 155.
132. Ibid.
133. Lacan, *The Four Fundamental Concepts of Psychoanalyisis*, p. 59.
134. By interpreting Berkeley's God as blind, we are going against his own understanding of God: "He is no Blind agent & in truth a blind Agent is a Contradiction." Berkeley, *Philosophical Commentaries*, p. 97, par. 812.
135. Beckett, "Film," p. 165.
136. Deleuze, "The Greatest Irish Film (Beckett's 'Film')," p. 23.
137. Ibid., p. 24.

Chapter 3 (Person)

1. Berkeley, "Philosophical Correspondence between Berkeley and Samuel Johnson," p. 417.
2. Ibid., p. 418.
3. Ibid., p. 426.
4. Ibid., p. 428.
5. Ibid., pp. 433–34.
6. Berkeley, *Philosophical Commentaries*, p. 22, par. 159.
7. Ibid.
8. Lacan, *The Seminar of Jacques Lacan, Book II*, p. 225.
9. Deleuze/Guattari, *Anti-Oedipus*, p. 26.
10. Berkeley, *Philosophical Commentaries*, p. 35, par. 282.
11. Ibid., p. 98, par. 823.
12. Berkeley, *Commonplace Book*, p. 112, par. 946.
13. Sillem, *George Berkeley and the Proofs for the Existence of God*, p. 162.
14. Ibid.
15. Berkeley, "De motu," p. 263, par. 33.
16. Derrida, *Memoirs of the Blind*, p. 65.
17. Berkeley, *The Principles of Human Knowledge*, pp. 102–3, pars. 75–76.
18. Preindividual singularities as sensible points, which inform the constellation, would also have to be the actualization of a potency.

19. Berkeley, *The Principles of Human Knowledge*, p. 67, par. 5.
20. Ibid., p. 77, par. 25.
21. Deleuze, "The Exhausted," p. 152.
22. Berkeley, "De motu," p. 260, par. 22.
23. Ibid., p. 275, par. 69.
24. Ibid., pp. 261–62, pars. 26–27.
25. Deleuze, "The Exhausted," p. 152.
26. Ibid., p. 155.
27. Ibid.
28. Ritchie, *George Berkeley*, p. 86.
29. Berkeley, *Philosophical Commentaries*, p. 14, par. 71.
30. Ritchie, *George Berkeley*, p. 86.
31. Berkeley, "Siris," p. 115, par. 240.
32. Ibid., p. 119, par. 249.
33. Berkeley, "De motu," p. 256, par. 5.
34. Ibid., p. 256, par. 4.
35. Berkeley, "Siris," p. 112, par. 234.
36. Ibid.
37. Ibid.
38. Berkeley, "De motu," p. 256, par. 4.
39. Ritchie, *George Berkeley*, p. 91.
40. Ibid., p. 91.
41. Ibid., p. 92.
42. Foucault, *The Order of Things*, p. 72.
43. Ibid.
44. Ritchie, *George Berkeley, A Reappraisal*, p. 91.
45. Deleuze, "Michel Tournier and the World Without Others," p. 307.
46. Ibid., p. 305.
47. Ibid., p. 310.
48. Ibid., p. 306.
49. Tournier, *Friday*, pp. 54–55.
50. Ibid., p. 55.
51. Deleuze, "Michel Tournier and the World Without Others," p. 306.
52. Tournier, *Friday*, p. 67.
53. Berkeley, "De motu," p. 275, par. 70.
54. Kant, *Critique of Pure Reason*, p. 162.
55. Ibid.: "The word community has in our language two meanings and contains the two notions conveyed in the Latin *communio* and *commercium*."
56. Derrida, *Given Time*, p. 82.
57. Ritchie, *George Berkeley, A Reappraisal*, pp. 84–85.
58. Ibid., p. 85.
59. Ibid.

60. Nancy, *The Sense of the World*, p. 61.
61. Ibid., p. 143.
62. Berkeley, *Philosophical Commentaries*, p. 98, par. 824: "My Doctrines rightly understood all that Philosophy of Epicurus, Hobbs, Spinoza, etc. which has been a Declared Enemy of Religion, Comes to ye ground."
63. Spinoza, *Ethics*, part 3, def. 1
64. Ibid., part 4, proposition 2, demonstration.
65. Ibid., part two, proposition 13, scholium.
66. Ibid., part four, proposition 4, corollary.
67. Berkeley, *Philosophical Commentaries*, pp. 98–99, par. 827.
68. Deleuze, "The Greatest Irish Film (Beckett's 'Film')," p. 26
69. Berkeley, *Philosophical Commentaries*, p. 98, par. 824.
70. Starobinski, *The Natural and Literary History of Bodily Sensation*, p. 354: "For a long time pain and pleasure were not ascribed to a particular sensor system."
71. Berkeley, *Philosophical Commentaries*, p. 39, par. 318.
72. Descartes, "The Passions of the Soul," p. 337, pars. 23, 24, 25.
73. Hegel, *Hegel's Philosophy of Mind*, p. 73, par. 400.
74. Berkeley, *Philosophical Commentaries*, p. 72, pars. 577, 578 and p. 76, par. 614.
75. Ibid., p. 72, par. 580.
76. Kant, *Critique of Pure Reason*, p. 96.
77. Deleuze, *Proust and Signs*, p. 105.
78. Berkeley, *Philosophical Commentaries*, p. 93, par. 775.
79. Deleuze, *Empiricism and Subjectivity*, p. 26.
80. Deleuze, *Difference and Repetition*, p. 77.
81. Deleuze, *Empiricism and Subjectivity*, p. 24.
82. Hume, *A Treatise of Human Nature*, p. 300.
83. Berkeley, *Philosophical Commentaries*, p. 97, par. 810.
84. Kant, *Critique of Pure Reason*, p. 163.
85. Hume, *A Treatise of Human Nature*, p. 301.
86. Descartes, "Meditation on First Philosophy," p. 21.
87. Berkeley, *Philosophical Commentaries*, p. 26, par. 201.
88. Hume, *A Treatise of Human Nature*, p. 301.
89. Deleuze, *Difference and Repetition*, p. 56.
90. Locke, *Essay on Human Understanding*, 1: 92.
91. See: Derrida, *Given Time*, p. 165: "It is worth recalling here how Kant situated man and radical evil in man: *between bestiality and diabolism*. . . . The order of the senses alone cannot explain this evil since sensibility deprives man of freedom and forbids one to speak of evil in this regard. By itself, sensibility would make of man an animal."
92. Ibid., p. 167.
93. Deleuze, *Difference and Repetition*, p. 151.
94. Ibid., p. 152.
95. Ibid., p. 148–49.

96. Pessoa, *The Book of Disquiet*, pp. 5–6.
97. Luce, *Berkeley and Malbranche*, p. 5.
98. Pessoa, *The Book of Disquiet*, p. 10.
99. Bozovic, "Berkeley: The Case of the Missing Spirit," p. 82.
100. Serres, *Genesis*, p. 30.
101. Berkeley, *Philosophical Commentaries*, p. 95, par. 791.
102. Pessoa, *The Book of Disquiet*, pp. 9–10.
103. Uspensky, *The Semiotics of the Russian Icon*, p. 38.
104. Pessoa, *The Book of Disquiet*, p. 10.
105. Serres, *Genesis*, p. 51.
106. Berkeley, *Philosophical Commentaries*, p. 82, par. 671.
107. Ibid., p. 53, par. 429.
108. Berkeley, *Philosophical Commentaries*, p. 90, par. 738.
109. Spinoza, *Ethics*, book II, proposition 21.
110. Borch-Jakobsen, "Who's Who?" p. 59.
111. Ibid.
112. Berkeley, *The Principles of Human Knowledge*, p. 138, par. 142.
113. Ibid., p. 136, par. 140: " that is we understand the meaning of the word."
114. Tipton, *Berkeley: The Philosophy of Immaterialism*, p. 270.
115. Berkeley, *The Principles of Human Knowledge*, p. 65, par. 1.
116. Ibid., introduction, p. 63, par. 25.
117. Berkeley, *Philosophical Commentaries*, p. 76, par. 615.
118. Ibid., p. 80, pars. 658–59.
119. Ibid., p. 81, par. 665.
120. Ibid., p. 81, par. 665 and p. 99, par. 828.
121. Ibid., p. 83, par. 681.
122. Ibid., p. 99, par. 828.
123. Ibid., p. 95, pars. 790–91 (emphasis added).
124. Deleuze, *Difference and Repetition*, p. 245.
125. Berkeley, *The Principles of Human Knowledge*, p. 65, par. 1.
126. Berkeley, *Philosophical Commentaries*, p. 9, par. 4.
127. Berkeley, *The Principles of Human Knowledge*, p. 113, par. 98.
128. Johnston, *The Development of Berkeley's Philosophy*, p. 241.
129. Berkeley, *Philosophical Commentaries*, p. 9, par. 13.
130. Tipton, *Berkeley: The Philosophy of Immaterialism*, p. 274.
131. Berkeley, *Philosophical Commentaries*, p. 9, par. 7.
132. Bergson, *Creative Evolution*, p. 12.
133. Berkeley, *The Principles of Human Knowledge*, p. 113, par. 98.
134. Berkeley, *Philosophical Commentaries*, p. 15, par. 83.
135. Deleuze, *Bergsonism*, p. 38.
136. Marcel Proust, "Contre Sainte-Beuve," p. 19.
137. Ibid.

138. See: Tipton, *Berkeley*, p. 273, and Cummins, "Perceptual Relativity and the Ideas in the Mind."

139. Florensky, *Iconostasis*, p. 41.

140. Ibid.

Chapter 3 (Eye)

1. Sartre, *Being and Nothingness*, p. 346.
2. Ibid.
3. Newton, "A New Theory About Light and Colours," p. 200.
4. Alpers, *The Art of Describing*, p. 36.
5. Berkeley, "An Essay Towards a New Theory of Vision," p. 39, par. 88. Although it should become self-evident, let us nevertheless note that we base our discussion on those "moments" of Berkeley's theory of vision that are directly connected with our theme—the relation of gaze and subjectivity. This means that we will not consider all the details of Berkeley's theory of vision and that we will not deal with the "physiology" or "psychology" of perception.
6. For the relation of immediate objects of vision to the inverted retinal image, see: Turbayne, "Berkeley and Molyneux on Retinal Images," pp. 339–55.
7. Berkeley, "An Essay Towards a New Theory of Vision," p. 41, par. 90.
8. Ibid., p. 51, par. 118.
9. In other words, Berkeley anticipated later insights that "up" and "down" are, in fact, only conventions. See the analysis of this problem in Merleau-Ponty, *The Phenomenology of Perception*, pp. 248–50.
10. Ritchie, *George Berkeley: A Reappraisal*, p. 20.
11. Ibid.
12. Derrida, *Specters of Marx*, pp. 100–101.
13. Berkeley, "An Essay Towards a New Theory of Vision," p. 41, par. 90.
14. Schwartz, *The Culture of the Copy*, p. 182.
15. Shakespeare, *Hamlet*, p. 37.
16. Kamuf, "Specters of Gender," p. 168. "It is all a question of a hauntology and not ontology, since we cannot speak of an ontology of the specter."
17. Derrida, *Specters of Marx*, p. 6.
18. Ibid.
19. Lacan, "Desire and the Interpretation of Desire in Hamlet," p. 52.
20. Berkeley, "An Essay Towards a New Theory of Vision," p. 41, par. 90.
21. Schwartz, *The Culture of the Copy*, p. 182.
22. Alpers, *The Art of Describing*, p. 34.
23. Ibid., p. 38.
24. Derrida, *Specters of Marx*, p. 7.
25. Alpers, *The Art of Describing*, p. 37.
26. Copjec, *Read My Desire*, p. 21.

27. Ibid., p. 35.
28. Atherton, *Berkeley's Revolution in Vision*, p. 149.
29. Berkeley, *Philosophical Commentaries*, p. 54, par. 438.
30. Pastore, *Selective History of Theories of Visual Perception*, p. 68.
31. Warnock, *Berkeley*, p. 27.
32. For analysis of these two conceptions of distance, see Brook, *Berkeley's Philosophy of Science*, pp. 65–74.
33. Atherton, *Berkeley's Revolution in Vision*, p. 73.
34. Warnock, *Berkeley*, p. 27.
35. In *Principles*, Berkeley explains that he wrote his *New Theory of Vision* in order to show "that distance or 'outness' is neither immediately of itself perceived by sight, nor yet apprehended or judged of by lines and angles." Berkeley, *Principles*, p. 84, par. 43.
36. Berkeley, "An Essay Towards a New Theory of Vision," p. 7, par. 2.
37. Ibid., p. 9, par. 11.
38. Uspensky, *The Semiotics of the Russian Icon*, p. 43.
39. Berkeley, *Philosophical Commentaries*, p. 34, par. 278: "Suppose inverting perspectives bound to ye eyes of a child, & continu'd to the years of Manhood, When he looks up or turns up his head he shall behold w^t we call under. Qu: w^t would he think of up & down?"
40. Uspensky, *The Semiotics of the Russian Icon*, p. 38.
41. Ibid., p. 44.
42. Ibid., p. 45.
43. Alpers, *The Art of Describing*, p. 133.
44. Uspensky, *The Semiotics of the Russian Icon*, p. 45.
45. For analysis of "Molyneux's problem" and Berkeley's "solution" to this problem, see Vesey, "Berkeley and the Man Born Blind," pp. 189–206 and Park, "Locke and Berkeley on the Molyneux Problem," pp. 253–60.
46. Diderot added another angle to Molyneux's problem—namely, that the question of whether Molyneux's man would succeed in distinguishing forms must be separated from the question of whether he would know how to name them. See Diderot, "Letter on the Blind, " p. 25.
47. See Pastore, *Selective History of Theories of Visual Perception*, p. 67: "Both Locke and Molyneux presuppose that the resighted man must have the opportunity of simultaneously seeing and feeling the objects so that he can later distinguish them on the basis of sight alone. . . . Their contemporary Synge, contested the negative answer. The two tactile ideas he [the person] has formed correspond to the characteristics of objects. . . . Synge concludes that by his sight alone he might be able to know which was the globe, and which the cube." Responding to Molyneux's question, Leibniz is closer to Synge than to Locke. He also refutes Locke's and Molyneux's negative answer in a way similar to Synge's, although he introduces into the argumentation the principle of reason: "If you will weigh my answer, you will find that I have put a condition in it which can be considered as included in the question; it is, that the only thing in question is

that of distinguishing, and that the blind man knows that the two figured bodies which he must distinguish are there, and that thus each of the appearances which he sees is that of the cube or that of the globe. In this case, it seems to me beyond doubt that the blind man who ceases to be blind can distinguish them by the principles of reasoning joined to what touch has provided him with beforehand of sensible knowledge." Leibniz, "Of Perception," p. 417.

48. In M. J. Morgan, *Molyneux's Question*, p. 19.
49. Ibid.
50. Berkeley, *Philosophical Commentaries*, p. 23, par. 174.
51. Ibid., p. 13, par. 62.
52. Atherton, *Berkeley's Revolution in Vision*, p. 75.
53. Armstrong, *Berkeley's Theory of Vision*, p. 5.
54. Merleau-Ponty, *The Phenomenology of Perception*, p. 255.
55. Locke, *Essay On Human Understanding*, 1:132.
56. Ibid.
57. Pastore, *Selective History*, p. 65.
58. Deleuze, *Cinema 1*, pp. 95–96.
59. Merleau-Ponty, *Phenomenology of Perception*, p. 216.
60. Zourabichvili, "Six Notes on the Percept," p. 190.
61. Deleuze, *Cinema 1*, p. 87.
62. Ibid., p. 89.
63. Ibid.
64. Locke, *Essay*, 1:132.
65. In Morgan, *Molyneux's Question*, p. 19.
66. Berkeley, *Philosophical Commentaries*, pp. 20 and 29, pars. 136 and 226.
67. Berkeley, "An Essay Towards a New Theory of Vision," p. 20, par. 43.
68. Mitchell, *Picture Theory*, p. 42.
69. Ibid., p. 48. Having in mind precisely this multistability, Wittgenstein claims the impossibility of naming the colors because there does not exist a generally accepted criterion of what color is "unless it is one of our colors": "But how do I know that I mean the same by the words 'primary colours' as some other person who is also inclined to call green a primary colour? No—here language-games decide." Wittgenstein, *Remarks on Colour*, pp. 2e–3e.
70. Deleuze, *Cinema 2*, p. 6.
71. Ibid., p. 7.
72. Merleau-Ponty, *Phenomenology of Perception*, p. 68.
73. Ibid.
74. Ibid.
75. Perhaps Berkeley's theory of vision anticipated the domination of the "empirical gaze" and knowledge that eigtheenth-century physicians had identified with "historical," as opposed to philosophical knoledge. See Foucault, *The Birth of the Clinic*, pp. 5 and 6: "The historical embraces whatever, *de facto* or *de jure*, sooner or later, directly or indi-

rectly may be offered to the gaze. . . . Disease is perceived fundamentally in a space of projection without depth, of coincidence without development." Of course, we are only referring to a similarity. Differences between Berkeley's theory of vision and the medical gaze of the late eighteenth century remain, nevertheless, irreducible. Among other reasons, this is because the clinical gaze implies that "the immediate on which it opens states the truth only if it is at the same time its origin, that is, its starting point, its principle and law of composition; and the gaze must restore as truth what was produced in accordance with a genesis: in other words, it must reproduce in its own operations what has been given in the very movement of composition. It is precisely in this sense that it is 'analytic.'" Ibid., p. 108. Berkeley's gaze, on the contrary, does not have it own genesis and cannot "develop" its own analytic.

76. Merleau-Ponty, *Phenomenology of Perception*, p. 283.

77. Ibid., p. 283. By referring to "spatiality without things," that is, to the "structure of the night," Merleau-Ponty introduces a paradox. He understands the night as "pure depth without foreground or background and without any distance separating it from me": "even the shouts or a distant light people it only vaguely, and then it comes to life in its entirety; it is pure depth." (Ibid.)

78. Ibid., p. 283.

79. Ibid., 262.

80. Becket, "Film," p. 169.

81. Deleuze, "The Greatest Irish Film (Beckett's 'Film')," p. 25.

82. Merleau-Ponty, *The Phenomenology of Perception*, p. 274.

83. Starobinski, *The Living Eye*, p. 2.

84. Ibid., p. 3.

85. Ibid.

86. Ibid.

87. Bryson, *Vision and Painting*, p. 94

88. See also Merleau-Ponty, *Phenomenology of Perception*, p. 27: "On the other hand, inattentive or delirious perception is a semi-torpor, describable only in terms of negations, its object has no consistency, the only objects about which one can speak being those of waking consciousness. It is true that we carry with us, in the shape of our body, an ever-present principle of absent-mindedness and bewilderment."

89. Bryson, *Vision and Painting*, p. 98.

90. Absence of spatial features from the visible lead some interpreters to claim that in Berkeley, strictly speaking, one cannot speak of the two-dimensionality of the picture because two-dimensionality also refers to the spatial organization of the visible. See Schwartz, *Vision*, p. 28.

91. Lyotard, "Figure Foreclosed," p. 84.

92. Lyotard, "Beyond Representation," p. 158.

93. Lyotard, "The Dream-Work Does Not Think," p. 29.

94. Berkeley, *Philosophical Commentaries*, p. 23, par. 165

95. Atherton, "How to Write the History of Vision," p. 156. See also p. 151: "Berkeley's theory is conditioned . . . by a notion of an innocent eye."
96. Atherton, *Berkeley's Revolution in Vision*, p. 99.
97. Berkeley, "An Essay Towards a New Theory of Vision," p. 28, par. 62.
98. Berkeley, *Philosophical Commentaries*, p. 29, par. 226.
99. Warnock, *Berkeley*, p. 42.
100. Luft, "Embodying the Eye of Humanism," p. 179.
101. Ibid., p. 180.
102. Vico, *The New Science*, p. 24.
103. Vico glimpses this logic of metaphorical writing while writing his own texts: "Most of Vico's metaphors are elementary. They are closely connected with nature and man as its part, and they themselves come from nature and are based on a broadening of the comparison with some natural phenomenon." See Roic, *Giambattista Vico*, p. 108.
104. Luft, "Embodying the Eye of Humanism," p. 176.
105. Berkeley, "An Essay Towards a New Theory of Vision," pp. 44–45, pars. 102–3.
106. Ibid., p. 22, par. 45.
107. Ibid., p. 26, par. 59.
108. Ibid., p. 23, par. 49.
109. Deleuze, *Logique de la sensation*, p. 99.
110. Deleuze/Guattari, *A Thousand Plateaus*, p. 494.
111. Deleuze, *Logique de la sensation*, p. 87.
112. Deleuze/Guattari, *A Thousand Plateaus*, p. 493.
113. Merleau-Ponty, *The Phenomenology of Perception*, p. 282.
114. Ibid., p. 283.
115. Warnock, *Berkeley*, p. 42.
116. Deleuze/Guattari, *Anti-Oedipus*, p. 131.
117. Ibid.
118. Derrida, "Roundtable on Translation," p. 106
119. Kamuf, "Singular Sense—Second Hand."
120. Derrida, "Roundtable on Translation," p. 101.

Works Cited

Alpers, Svetlana. *The Art of Describing*. Chicago: Univeristy of Chicago Press, 1984.
Armstrong, D. M. *Berkeley's Theory of Vision*. Melbourne: Melbourne University Press, 1960.
Artaud, Antonin. "Van Gogh: The Man Suicided by Society." In *Selected Writings*. Ed. Susan Sontag. Berkeley: University of California Press, 1988.
Atherton, Margaret. *Berkeley's Revolution in Vision*. Ithaca: Cornell University Press, 1990.
———. "How to Write the History of Vision: Understanding the Relationship Between Berkeley and Descartes." In *Sites of Vision*. Ed. D. M. Levin. Cambridge, Mass.: MIT Press, 1997.
Badiou, Alain. "The Fold: Leibniz and the Baroque." In *Gilles Deleuze and the Theater of Philosophy*. Ed. C. V. Boundas and D. Olkowski. Trans. Thelma Sowley. New York: Routledge, 1994.
Barthes, Roland. "On Photography." In *The Grain of the Voice*. Trans. Linda Coverdale. Berkeley: University of California Press, 1985.
Beckett, Samuel. "Film." In *Collected Shorter Plays*. New York: Grove Press, 1984.
Bergson, Henri. *Creative Evolution*. Trans. Arthur Mitchell. Mineola, N.Y: Dover, 1998.
Berkeley, George. "Alciphron, or the Minute Philosopher." In *The Works of George Berkeley Bishop of Cloyne*. Vol. 3. Ed. A. A. Luce and T. E. Jessop. London: Thomas Nelson & Sons, 1964.
———. *Commonplace Book*. (*Philosophical Commentaries*). Ed. G. A. Johnston. London: Faber and Faber, 1930.
———. "The Eruption of Mount Vesuvius." In *The Works of George Berkeley Bishop of Cloyne*. Vol. 4, *Writings on Natural History*. Ed. A. A. Luce and T. E. Jessop. London: Thomas Nelson and Sons, 1964. Reprinted from 1953 edition.
———. "An Essay Towards a New Theory of Vision." In *Philosophical Works*. Ed. Michael R. Ayers. London: J. M. Dent, 1975.
———. "De motu." In *Philosophical Works*.
———. *Philosophical Commentaries*. In *The Works of George Berkeley Bishop of Cloyne*. Vol. 1. Ed. A. A. Luce and T. E. Jessop. London: Thomas Nelson and Sons, 1964.

———. "Philosophical Correspondence between Berkeley and Samuel Johnson." In *Philosophical Works*.

———. *The Principles of Human Knowledge*. Ed. G. J. Warnock. Glasgow: William Collins Sons, 1977.

———. "Siris: A Chain of Philosophical Reflextions and Inquiries." In *The Works of George Berkeley Bishop of Cloyne*. Vol. 5. Ed. A. A. Luce and T. E. Jessop. London: Thomas Nelson and Sons, 1964.

———. "The Theory of Vision Vindicated and Explained." In *Philosophical Works*.

Blumenberg, Hans. "Light as a Metaphor for Truth." In *Modernity and the Hegemony of Vision*. Ed. D. M. Levin. Berkeley: University of California Press, 1993.

Borch-Jakobsen, Mikkel. "Who's Who? Introducing Multiple Personality." In *Supposing the Subject*. Ed. Joan Copjec. London: Verso, 1994.

Bozovic, Miran. "Berkeley: The Case of the Missing Spirit." *Razpol 5*. Ljubljana: Drustvo za teoretsko psihoanalizo, 1989.

Brook, Richard J. *Berkeley's Philosophy of Science*. The Hague: Martinus Nijhoff, 1973.

Bruno, Giordano, *Cause, Principle and Unity*. Ed. R. J. Blackwell and R. de Lucca. Trans. R. de Lucca. Cambridge: Cambridge University Press, 1989.

———. "Essays on Magic." In *Cause, Principle and Unity*.

———. *The Heroic Frenzies*. Trans. Paul Eugene Memmo, Jr. Chapel Hill: University of North Carolina Press, 1964 (www.esotericarchives.com/bruno/furori.htm).

Bryson, Norman. *Vision and Painting: The Logic of the Gaze*. London: Macmillan, 1983.

Burnyeat, M. F. "Idealism and Greek Philosophy: What Descartes Saw and Berkeley Missed." In *Idealism Past and Present*. Ed. Godfrey Vesey. Cambridge: Cambridge University Press, 1982.

Calderon de la Barca, Pedro. *Life is a Dream*. In *Six Plays*. Trans. E. Honig. New York: Iasta Press, 1993.

Copjec, Joan. *Read My Desire: Lacan Against the Historicists*. Cambridge, Mass.: MIT Press, 1995.

Cummins, P. D. "Perceptual Relativity and the Ideas in the Mind." *Philosophy and Phenomenological Research* 24 (1963).

Damisch, Hubert. *The Origin of Perspective*. Cambridge, Mass.: MIT Press, 1995.

Dancy, J. *Berkeley: An Introduction*. Oxford: Basil Blackwell, 1987.

Deleuze, Gilles. *Bergsonism*. Trans. Hugh Tomlinson and Barbara Habberjam. New York: Zone Books, 1991.

———. *Cinema 1: The Movement-Image*. Trans. Hugh Tomlinson and Barbara Habberjam. Minneapolis: University of Minnesota Press, 1997.

———. *Cinema 2: The Time-Image*. Trans. Hugh Tomlinson and Robert Galeta. Minneapolis: University of Minnesota Press, 1989.

———. *Difference and Repetition*. Trans. Paul Patton. New York: Columbia University Press, 1994.

———. *Empiricism and Subjectivity: An Essay on Hume's Theory of Human Nature*. Trans. Constantin V. Boundas. New York: Columbia University Press, 1991.

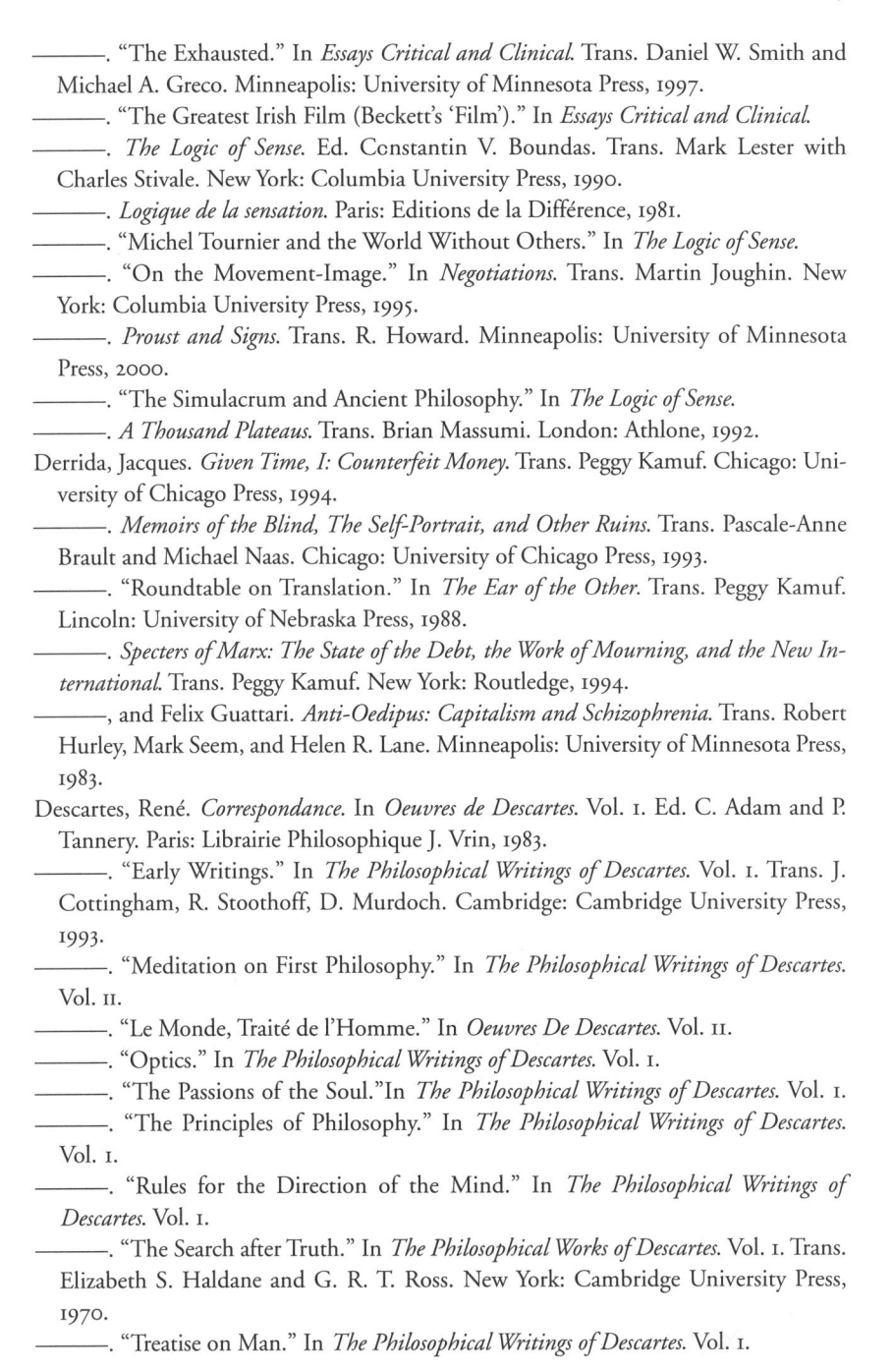

———. "The Exhausted." In *Essays Critical and Clinical.* Trans. Daniel W. Smith and Michael A. Greco. Minneapolis: University of Minnesota Press, 1997.

———. "The Greatest Irish Film (Beckett's 'Film')." In *Essays Critical and Clinical.*

———. *The Logic of Sense.* Ed. Constantin V. Boundas. Trans. Mark Lester with Charles Stivale. New York: Columbia University Press, 1990.

———. *Logique de la sensation.* Paris: Editions de la Différence, 1981.

———. "Michel Tournier and the World Without Others." In *The Logic of Sense.*

———. "On the Movement-Image." In *Negotiations.* Trans. Martin Joughin. New York: Columbia University Press, 1995.

———. *Proust and Signs.* Trans. R. Howard. Minneapolis: University of Minnesota Press, 2000.

———. "The Simulacrum and Ancient Philosophy." In *The Logic of Sense.*

———. *A Thousand Plateaus.* Trans. Brian Massumi. London: Athlone, 1992.

Derrida, Jacques. *Given Time, I: Counterfeit Money.* Trans. Peggy Kamuf. Chicago: University of Chicago Press, 1994.

———. *Memoirs of the Blind, The Self-Portrait, and Other Ruins.* Trans. Pascale-Anne Brault and Michael Naas. Chicago: University of Chicago Press, 1993.

———. "Roundtable on Translation." In *The Ear of the Other.* Trans. Peggy Kamuf. Lincoln: University of Nebraska Press, 1988.

———. *Specters of Marx: The State of the Debt, the Work of Mourning, and the New International.* Trans. Peggy Kamuf. New York: Routledge, 1994.

———, and Felix Guattari. *Anti-Oedipus: Capitalism and Schizophrenia.* Trans. Robert Hurley, Mark Seem, and Helen R. Lane. Minneapolis: University of Minnesota Press, 1983.

Descartes, René. *Correspondance.* In *Oeuvres de Descartes.* Vol. 1. Ed. C. Adam and P. Tannery. Paris: Librairie Philosophique J. Vrin, 1983.

———. "Early Writings." In *The Philosophical Writings of Descartes.* Vol. 1. Trans. J. Cottingham, R. Stoothoff, D. Murdoch. Cambridge: Cambridge University Press, 1993.

———. "Meditation on First Philosophy." In *The Philosophical Writings of Descartes.* Vol. II.

———. "Le Monde, Traité de l'Homme." In *Oeuvres De Descartes.* Vol. II.

———. "Optics." In *The Philosophical Writings of Descartes.* Vol. 1.

———. "The Passions of the Soul." In *The Philosophical Writings of Descartes.* Vol. 1.

———. "The Principles of Philosophy." In *The Philosophical Writings of Descartes.* Vol. 1.

———. "Rules for the Direction of the Mind." In *The Philosophical Writings of Descartes.* Vol. 1.

———. "The Search after Truth." In *The Philosophical Works of Descartes.* Vol. 1. Trans. Elizabeth S. Haldane and G. R. T. Ross. New York: Cambridge University Press, 1970.

———. "Treatise on Man." In *The Philosophical Writings of Descartes.* Vol. 1.

———. "The World, or Treatise on Light." In *The Philosophical Writings of Descartes.* Vol. 1.

Diderot, Denis. "Letter on the Blind." In *Diderot's Selected Writings.* Ed. L. G. Crocker. Trans. Derek Coltman. New York: Macmillan, 1966.

Florensky, Pavel. *Iconostasis.* Tran. Donald Sheehan and Olga Andrejev. New York: St. Vladimir's Seminary Press, 2000.

Foucault, Michel. *The Birth of the Clinic: An Archaeology of Medical Perception.* Trans. A. M. Sheridan Smith. New York: Vintage, 1994.

———. *Death and the Labyrinth: The World of Raymond Roussel.* Trans. Charles Ruas. Berkeley: University of California Press, 1986.

———. *The Order of Things: An Archaeology of the Human Sciences.* New York: Vintage, 1994.

———. *This Is Not a Pipe.* Trans. James Harkness. Berkeley: University of California Press, 1983.

———. *This is Not a Pipe: Aesthetics, Method, and Epistemology.* Ed. J. D. Faubion. Penguin, 1998.

Godwin, Joscelyn. *Athanasius Kircher: A Renaissance Man and the Quest for Lost Knowledge.* London: Thames and Hudson, 1979.

Goux, Jean-Joseph. "Descartes et la perspective." *L'Esprit Créateur* 25, no. 1 (1985).

Hegel, G.W. F. *Philosophy of Mind.* Part 3 of the Encyclopedia of the Philosophical Sciences. Trans. William Wallace. Oxford: Clarendon Press, 1971.

Hocke, Gustav Rene. *Manierismus in der literatur: Sprach-Alchemie und esoterische Kombinationskunst.* Reinbek bei Hamburg: Rowohlt, 1957.

Hume, David. *A Treatise of Human Nature.* Ed. E. C. Mossner. London: Penguin, 1987.

Johnston, G. A. *The Development of Berkeley's Philosophy.* London: Methuen, 1923.

Kamuf, Peggy. "Singular Sense—Second Hand." Unpublished manuscript.

———. "Specters of Gender." *Zenske Studije* (Women's studies) 10, Belgrade (1998).

Kant, Immanuel. *Critique of Pure Reason.* Trans. J. M. D. Meiklejohn. London: J. M. Dent and Sons, 1956.

Krauss, Rosalind E. *The Optical Unconscious.* Cambridge, Mass.: MIT Press, 1994.

Lacan, Jacques. "Desire and the Interpretation of Desire in Hamlet." In *Literature and Psychoanalysis.* Ed. Shoshana Felman. Baltimore: The Johns Hopkins University Press, 1982.

———. *The Four Fundamental Concepts of Psychoanalysis.* Ed. J.-A. Miller. Trans. Alan Sheridan. New York: Norton, 1998.

———. *The Seminar of Jacques Lacan: Book I, Freud's Papers on Technique 1953–1954.* Ed. J.-A. Miller. Trans. John Forrester. Cambridge: Cambridge University Press, 1988.

———. *The Seminar of Jacques Lacan: Book II, The Ego in Freud's Theory and in the Technique of Psychoanalysis, 1954–1955.* Ed. J.-A. Miller. Trans. Sylvana Tomaselli. Cambridge: Cambridge University Press, 1988.

Leibniz, Gottfried Wilhelm. "An Example of Demonstrations about the Nature of Corporeal Things, Drawn from Phenomena." In *Philosophical Papers and Letters*. Ed. Leroy E. Loemker. Dordrecht: D. Reidel, 1976.

———. "Letter to Jacob Thomasius, April 20/30, 1669." In *Philosophical Papers and Letters*.

———. "The Monadology." In *G. W. Leibniz: Selections*. Ed. Philip P. Wiener. New York: Charles Scribner's Sons, 1951.

———. "Of Perception." In *G. W. Leibniz: Selections*.

———. "Tentamen Anagogicum: An Anagogical Essay in The Investigation of Causes." In *Philosophical Papers and Letters*. Locke John. *Essay on Human Understanding. The Works of John Locke*. 10 volumes. London: Thomas Tegg; W. Sharpe and Son, 1823. Reprinted by Scientia Verlag Aalen, Germany, 1963.

Lowry, Malcolm. *Under The Volcano*. New York: Penguin, 1971.

Luce, A. A. *Berkeley and Malebranche: A Study in the Origins of Berkeley's Thought*. London: Oxford University Press, 1967.

Luft, Sandra Rudnick. "Embodying the Eye of Humanism: Giambattista Vico and the Eye of Ingenium." In *Sites of Vision*. Cambridge, Mass.: MIT Press, 1997.

Lyotard, Jean-Francois. "Acinema." In *The Lyotard Reader*. Ed. A. Benjamin. Oxford: Basil Blackwell, 1989.

———. "Beyond Representation." In *The Lyotard Reader*.

———. "The Dream-Work Does Not Think." In *The Lyotard Reader*.

———. "Figure Foreclosed." In *The Lyotard Reader*.

———. "Scapeland." In *The Inhuman*. Trans. Geoffrey Bennington and Rachel Bowlby. Cambridge: Polity, 1991.

Merleau-Ponty, Maurice. "Eye and Mind." In *The Merleau-Ponty Aesthetics Reader*. Ed. and Trans. Michael B. Smith. Evanston, Ill.: Northwestern University Press, 1993.

———. *The Phenomenology of Perception*. Trans. Colin Smith. New York: Routledge, 2000.

———. *The Visible and the Invisible*. Trans. Alphonso Lingis. Evanston, Ill.: Northwestern University Press, 1997.

Mitchell, W. J. T. *Iconology*. Chicago: University of Chicago Press, 1986.

———. *Picture Theory: Essays on Verbal and Visual Representation*. Chicago: University of Chicago Press, 1994.

Morgan, M. J. *Molyneux's Question: Vision, Touch and the Philosophy of Perception*. Cambridge: Cambridge University Press, 1977.

Nancy, Jean-Luc. *The Sense of the World*. Trans. J. S. Librett. Minneapolis: University of Minnesota Press, 1997.

Newton Isaac. "A New Theory about Light and Colours." In *English Science Bacon to Newton*. Ed. Brian Vickers. Cambridge: Cambridge University Press, 1987.

Park, Desirée. "Locke and Berkeley on the Molyneux Problem." *Journal of the History of Ideas* 30 (1969).

Pascal, Blaise. "De l'Esprit géometrique." In *Oeuvres Completès*, Paris, Bibl. de la Pléiade 1964.

———. *Pensées*. Trans. Honor Levi. Oxford: Oxford University Press, 1995.
Pastore, Nicholas. *Selective History of Theories of Visual Perception, 1650–1950*. New York: Oxford University Press, 1971.
Pessoa, Fernando. *The Book of Disquiet*. London: Pantheon, 1991.
Pitcher, George. *Berkeley*. London: Routledge and Kegan Paul, 1977.
Proust, Marcel. "Contre Sainte-Beuve." In *Marcel Proust on Art and Literature*. Trans. Sylvia Townsend Warner. New York: Carroll and Graf, 1997.
Ritchie, A. D. *George Berkeley: A Reappraisal*. Manchester: Manchester University Press, 1967.
Roic, Sanja. *Giambattista Vico: Knizevnost, retorika, poetika* (Giambatista Vico: Literature, Rhetoric, Poetics). Zagreb: Hrvatsko filozofsko drustvo, 1990.
Rorty, Richard. *Philosophy and the Mirror of Nature*. Princeton: Princeton University Press, 1979.
Sabra, A. I. *Theories of Light from Descartes to Newton*. Cambridge, 1987.
Sartre, Jean-Paul. *Being and Nothingness*. Trans. Hazel E. Barnes. New York: Washington Square Press, 1956.
Schelling, F. W. J. *Bruno; or, On the Natural and the Divine Principle of Things*. Ed. and Trans. Michael G. Vater. Albany: SUNY University Press, 1984.
Schwartz, Hillel. *The Culture of the Copy*. New York: Zone Books, 1996.
Schwartz, Robert. *Vision: Variations on Some Berkeleian Themes*. Oxford : Basil Blackwell, 1994.
Serres, Michael. *Genesis*. Trans. Genevieve James and James Nielson. Michigan: University of Michigan Press, 1995.
———. "Panoptic Theory." In *The Limits of Theory*. Ed. Thomas M. Kavanagh. Stanford: Stanford University Press, 1989.
Shakespeare, William. *Hamlet*. London: Penguin, 1994.
Sillem, Edward A. *George Berkeley and the Proofs for the Existence of God*. London: Longmans, Green, 1975.
Spinoza, Baruch de. *Ethics*. Ed. and trans. G. H. R. Parkinson. Oxford: Oxford University Press, 2000.
Starobinski Jean. *The Living Eye*. Cambridge, Mass.: Harvard University Press, 1989.
———. *The Natural and Literary History of Bodily Sensation*. In *Fragments for a History of the Human Body*, Part Two. Ed. Michel Feher, Ramona Naddaff, and Nadia Tazi. Trans. Ralph Manheim. Pp. 351–405. New York: Zone 4, 1989
Swift, Jonathan. *Gulliver's Travels*. Penguin, 1994.
Tipton, I. C. *Berkeley: The Philosophy of Immaterialism*. London: Methuen, 1974.
Tournier, Michel. *Friday*. Trans. Norman Denny. Baltimore: The Johns Hopkins University Press, 1997.
Turbayne, C. M. "Berkeley and Molyneux on Retinal Images." *Journal of the History of Ideas* 16 (1955).
Uspensky, Boris. *The Semiotics of the Russian Icon*. Ed. Stephen Rudy. Peter De Ridder, 1976.

Vesey, N. A. "Berkeley and the Man Born Blind." *Proceedings of the Aristotelian Society* 61 (1960–61).
Vico, Giambattista. *The New Science.* Trans. David Marsh. London: Penguin, 1999.
Warnock, G. J. *Berkeley.* Oxford: Basil Blackwell, 1982.
Wilson, Catherine. "Discourses of Vision in Seventeenth Century Metaphysics." In *Sites of Vision: The Discursive Construction of Sight in the History of Philosophy.* Ed. David Michael Levin. Cambridge, Mass.: MIT Press, 1997.
Wittgenstein, Ludwig. *Remarks on Colour.* Ed. G. E. M. Anscombe. Trans. Linda L. McAlister and Margarete Schättle. Berkeley: University of California Press, 1977
Yates, Frances A. *The Art of Memory.* Chicago: University of Chicago Press, 1974.
Zourabichvili, François. "Six Notes on the Percept." In *Deleuze: A Critical Reader.* Ed. Paul Patton. Cambridge: Blackwell, 1996.

Index

abstraction: and abstract ideas, 114; and adequacy, 81; and the binocular image, 33; and central perspective, 154; and the eye of ingenium, 170; and the internal spectator, 154; and motion, 108; and the proper objects of sight, 161; and self-knowledge, 127; and the visible, 44
animal, 4, 6, 18–19, 26–29, 32–33, 46, 66, 72, 74–77, 80, 108, 126–27
archetype, 97–100

Babylon, 170, 174
Bacon, Roger, 146–47
Barthes, Roland, 60
Beckett, Samuel, 49–50, 93, 118, 163; and self-perception, 86; and god's eye, 94; and exhaustion, 107; and anonymous person, 159
Bergson, Henri, 128, 135
Berkeley, George, 48, 70, 95, 97; and abstract ideas, 77–82; and the archetype, 98–103; and apperception, 124–25; and the body 64, 115–17, 120; and Cartesian optics 54–56; and chaos, 74; and distance 92–94, 152–53; and intuition, 130; and objects of vision, 50–52; and solipsism, 109–15; and time, 135; and the virtual 85–86, 116
blindness: and allegory, 12; and apperception, 95; and copula, 13; and differentiation, 11; and inverted time, 137–38; and madness, 41–44; and perspective, 38–39; and projection, 30–32; and resemblance, 15–16; and similarity, 13–14;
Bond, James, 58
Bruno, Giordano, 1, 5, 7, 10, 11–12, 13–17

calligram, 59–60
close-up, 159, 160, 164

cogito, 34, 37, 41–44, 130, 148
constats, 162
contemplation, 8, 10–11, 13–14, 17, 22–23, 122–23

Defoe, Daniel: and Robinson Crusoe 109–12, 140
Derrida, Jacques, 150
Descartes, René, 18; and the blind man, 32–33; and cat's eye, 22; and the cerebral cortex, 86; and the copper engraving, 25–27; and distance, 151–58; and the evil demon, 147; and figure, 160; and geometrical projections, 51–52; and geometrical space, 32; and Kepler, 149; and Newton, 140–42; and painting, 42; and the point of the subject, 42; and recognition, 124; and the reflexive gaze 22, 40; and sensations, 121; and the will, 133
desire, 10–14, 22, 35, 50, 74, 76–77, 100–105, 117–19, 127, 138

Florensky, Pavel, 91, 138–39

gaze: and aspect, 89; and contemplation, 3–11; and glance 149, 165–67; and god 86–87, 89–91, 93; and the icon, 91; and the locked room, 72; and perspective, 30, 34–36, 137, 145; and the retinal image, 148; and res cogitans, 36–38, 41; and spectacula, 146; and suggestion, 59; and touch, 169–72

Hamlet, 1, 147
Hegel, G. W. F., 120
Heller, Joseph: and *Catch 22*, 132–33
Holmes, Sherlock, 57–58
Hume, David, 110, 123–25

209

iconography, 91–96, 114, 139, 163–64; and the iconographic subject, 129–31, 134, 139, 153, 159, 163–64, 174; and the Byzantine icon, 93

Johnson, Samuel, 97–99, 102

Kant, Immanuel, 116, 122, 126, 134
Kepler, Johannes, 143–44, 149–52, 154
Klee, Paul, 61

Lacan, Jacques, 56
Leibniz, G. W., 78–84, 92–94

madness: and the blind spot of perspective, 41–42; and the frenzied eye, 10; and geometrical vision, 22; and identity, 86; and *res cogitans*, 47; and schizophrenia, 173–74
metaphor: and the audible, 169–70; and deduction 3, and the mirror, 4–5; as oppositional metaphor, 9; and picture, 54; and theater, 125–26; and the trope of vision, 27; and the universe, 2
mirror: and dreaming, 47; and the eye, 163; and god, 4; and image, 86, 145, 151; as linguistic figure 2–3; and passivity, 5–7; and subjectivity, 3; and visual perception, 52; and the world, 9
metapictures, 161
Molyneux's question, 152–53, 156, 160

Newton, Isaac, 134, 141–42

panopticon, 95
Pascal, Blaise, 42
passivity, 1, 5, 7, 100–107, 116–20; and exhaustion, 105, 118, 123; and joy, 15, 27, 57, 118–120, 172; and pain, 15, 63, 117–19, 135, 172, 174; and tiredness, 107; and suffering 115, 117–18, 135, 174
percepts, 125, 159
perspective: and depth, 72, 87, 159; and finitude, 5, 37; and foreground, 114; and the geometrical point, 34–35; and the haptic, 173; as horizontal perspective, 88–89; and interiority, 129; as inverted perspective, 90–91, 137–41, 154–62; and meditations, 41; and Molyneux's question, 153; and nocturnal space, 163; and optical space, 71; and the point of the gaze, 137; as *prospettiva accelerata*, 6; as *prospettiva rallentata*, 6; and scenography, 36, 89, 92; and the subject, 37–40
photograph, 45, 60
Plato, 4, 44
Proust, Marcel, 137

Sartre, Jean-Paul, 140–41
scenography, 36, 47–48, 87–92
simulacra, 56–59; 62–65; 159; and copies, 52–54, 56, 99
Spinoza, Baruch de, 77, 105, 119–20, 129

touch: and blindness, 36; and Cartesian optics, 32–33; and distance, 167; and the gaze, 168–71; and haptic space, 173; and ideas, 171–72; and the innocent eye, 156; and Molyneux's question, 161; and the position of objects, 144; and the proper objects of touch, 51; and sensation, 64; and suffering, 115–24
Tournier, Michel, 112

Vico, Giambattista, 169